I0485665

Antennas as Transmission Lines:

Linear Antenna Analysis and Design Requiring Only Algebra and Trigonometry

By John Lenahan

KØRW

Copyright © 2022, John C. Lenahan, Jr.

All rights reserved

Except as permitted under U.S. Copyright Law, no part of this book may be reprinted, reproduced, transmitted, or utilized in any form by any electronic, mechanical, or other means, now known or hereafter invented, including photocopying, microfilming, and recording, or in any information storage or retrieval system, without written permission of the author.

The author has attempted to trace the copyright holders of all material reproduced in this publication and apologizes to copyright holders if permission to publish in this form has not been obtained. If any copyright material has not been acknowledged please email booksbyrw@gmail.com and let me know so I may rectify in any future reprint.

ISBN: 978-1-387-85367-0

Contents

Chapter 1 Linear Antennas as Transmission Lines *1*

1.1 Models, Approximations, Accuracy and Precision *2*

1.2 Some Basic Transmission Line Theory *5*

1.3 Antenna Characteristic Impedance *8*

Chapter 2 Inductive Loading Short Antennas *14*

2.1 The Short Antenna and Feed Point Inductive Loading *16*

2.2 Antenna Reactance, Resistance, Bandwidth and "Q" *19*

2.3 Center vs Feed Point Inductive Loading *22*

2.4 Center Loading of Different Diameter Sections *26*

2.5 The Shorted Transmission Line Inductive "Stub" *28*

Chapter 3 Capacitive Loading Short Antennas *30*

3.1 Capacitive End Loading - The "Hat" *31*

3.2 The Pizza Pan Hat – Disk and Inductive End Loading *35*

3.3 The Two-Band Pizza Pan Monopole *36*

3.4 Capacitive End Loading with One Horizontal Wire - The "Inverted L" *38*

3.5 Capacitive End Loading with Two Horizontal Wires - The "T" Antenna *42*

3.6 Capacitive End Loading Using an Open Transmission Line "Stub" *45*

3.7 The 1/4-Wave Monopole - An Impedance Transformer *47*

3.8 The Longer Than 1/4-Wave Monopole - A Two-Band Antenna   *50*

3.9 A Three-Band Monopole *52*

3.10 A 50 Ohm Radiation Resistance Monopole *54*

Chapter 4 Monopoles Do Not Require Perfect Ground   *56*

4.1 Using Capacitive Bottom Loading – The Counterpoise   *57*

4.2 A Capacitive Bottom Loaded Three-Band Monopole   *59*

4.3 Reducing Counterpoise Length Using Inductive Loading *62*

4.4 Capacitive End Loading the 50 Ohm Counterpoise Monopole   *64*

4.5 An End Loaded, Bottom Counterpoise, 50 Ohm Monopole   *65*

4.6 The Bottoms-Up Counterpoise 50 Ohm Monopole *70*

4.7 An Inductive Counterpoise   *74*

Chapter 5 The Horizontal Dipole   *76*

5.1 The Dipole, Two Monopoles in Series – The Image Becomes Real   *77*

5.2 The Horizontal Dipole Meets Ground *80*

5.3 The Inverted V Dipole   *86*

5.4 The Harmonically Operated Dipole   *89*

5.5 The Off-Center Fed Horizontal Dipole   *94*

5.6 An Inductively Loaded Horizontal Dipole   *100*

5.7 A Two Frequency Antenna Using a "Trap"   *103*

5.8 The Bent Ends Horizontal Dipole   *112*

5.9 The Maximum Gain Dipole   *115*

Chapter 6 Vertical Dipoles and the Vertically Polarized Loop   *121*

6.1 The Vertical Dipole   *121*

6.2 An Octave Vertical Dipole   *124*

6.3 A Capacitive End Loaded Vertical Dipole   *127*

6.4 Vertical Dipole Asymmetrical Capacitive End Loading   *130*

6.5 The Flagpole Loophole Vertical Dipole   *134*

6.6 Low Vertical Loops   *138*

Appendix

A1 Afterwords and Afterthoughts   *143*

A2 Free Space – The Transmission Line of the Universe   *152*

A3 Antenna Gain – A Chewing Gum Theorem   *154*

A4 Formula List   *156*

Chapter 1 Linear Antennas As Transmission Lines

These first three sections lay a foundation for what is to follow in all later chapters.

The first section gives background for working with calculated values. The simple act of finding the resonant frequency length of a monopole or dipole requires a calculation which can result in numbers that are often rounded in application.

The second section is a discussion of transmission line theory in preparation for the third section. The concept of how different lengths of open or shorted transmission lines can act as capacitors, inductors and tuned circuits is introduced. This transmission line information sets up the use of transmission line theory for application to linear antennas.

Section 3 introduces a formula for finding the average equivalent characteristic impedance of the monopole antenna as a transmission line model. With this equivalent characteristic impedance, and transmission line theory from section 2, we can find the reactive part of a non-resonant antenna impedance.

Also in section 3 the reactance values calculated using the antenna as a transmission line model are compared to software models. A formula will be provided to find the estimated radiation resistance of a short monopole. The result is that by viewing an antenna as a transmission line we can calculate reactance and resistance at the monopole feed point to form a series AC circuit model for comparison to software models and later, to measured values of some real antennas.

The use of transmission line theory to characterize an antenna was once common and showed good agreement with measurements taken at the antenna.

Another method of analyzing an antenna by mathematically chopping it into little bits and then summing those pieces into a whole was developed. This "method of moments" (MoM) technique involved the use of calculus which was very calculation intensive and little used by engineers. It was not until digital computers with sufficient power and speed came along that the transmission line method was replaced by the method of moments.

MoM software will be used to validate the older transmission line model. Before software, the only way to validate the transmission line model was by measuring constructed scale models or full size antennas.

You will see, as you move through the sections, I am analyzing, designing and modifying antennas using the transmission line model. You become a participant in antennas and their design, rather than a spectator.

Having worked with the transmission line model so much, for linear antennas, in some cases, I can calculate AC circuit values faster than I can fire-up the software and enter the conductor coordinates. I do use the software to check my calculations but mostly I use the software to find radiation patterns, gains and elevation angles of the radiated field. The validation of the transmission line model by MoM now makes it quite possible to design your own antennas.

As much as I enjoy the power of the computer software, its convenience has taken away from a fundamental understanding of antennas.

What follows, throughout this book, are examples of possible real life antennas modeled as AC circuits using transmission line methods, then checked by software modeling. In later sections I construct examples of designed antennas and have measured results to compare with the transmission line model and the software model.

What I hope you will find, on this adventure, is that the older transmission line model can change how you see antennas by thinking of them as transmission lines to be solved using basic AC theory.

In a few of the examples in this book I have included the software line models. As you go through the examples I suggest you make sketches of the antennas for yourself. I feel this is part of the thinking about antennas process. Antenna software uses the X,Y,Z, coordinate system and your sketches will get you into thinking in 3-D.

"antennas have always behaved in strict accordance with circuit or transmission-line rules"
Les Moxon, G6XN, "HF Antennas For All Locations", 2nd Edition

1.1 Models, Approximations, Accuracy and Precision

Models

The AC circuit models of antennas developed using transmission line theory in this book will be approximations for ideal conditions. The ideal model for a monopole is over a perfectly conducting infinite ground plane, free space for dipoles.

After the ideal antenna as transmission line model is developed and applied to several examples, consideration of the effects for non-ideal conditions will be explored, producing a higher level approximation.

Approximations

Approximations are often byproducts of calculations. As an example, Pi (π) is often approximated to some number of digits which is not its exact value.

So what are we to do with calculated values that force us into the world of approximate numbers?

There are rules that involve the use of significant digits. One of the most basic "rounding" rules is not to use a calculated number of digits greater than your instruments can measure. Having a calculated value to 6-digits when your instrument can give only 3-digits is unreasonable, so we round.

In this book, I chose not to report final results of calculations beyond 3-digits due to the limitation of my instrument resolution. In the case of calculated values of angles, I will record them to 4-digits.

Precision

Precision has to do with how small a part of something can be measured. This is sometimes called resolution. An example might be a frequency counter that can measure to a few parts per million (PPM), or a meter's ability to measure small parts of Ohms, Volts or Amps.

Accuracy

Accuracy and precision are often confused or thought to be "almost" the same. They are not. If you are not precise in your work, you can produce inaccurate results. Accuracy is the relationship of a measured amount to some known, and agreed upon, standard. How close is that wooden meter stick length to the atomically derived standard meter?

An instrument that is precise, that has high resolution, through neglect or abuse, can become quite inaccurate. By reading the instrument manual you will see accuracy and precision are expressed differently.

As an example of how precision does not equal accuracy, here is a comparison of two different antenna software results for the feed point resistance and reactance of a 10ft conduit as a monopole over a perfect, infinite ground plane.

The calculated resonant frequency is, 234/10ft = 23.4MHz.

	Software #1	Software#2
23.4MHz	35.89+j0.425	35.82645+j1.203265
14.1MHz	8.467-j205.8	8.485128-j209.3666
18.1Mhz	16.13-j107.5	16.21379-j109.9392
21.1MHz	25.42-j45.83	25.46508-j47.25796
24.9MHz	45.05+j29.55	45.85715+j29.1377
28.1MHz	75.75+j99.79	74.92778+j99.96261

Which is more accurate? Software #2 might be considered by some as "more accurate" because it has more digits. But, as we now know, accuracy and precision are not the same thing.

Truth be told, both software versions have calculated approximate values based on their own virtual antenna model. If we round the results to 3-digits (as proposed), there is very little difference between the two software calculated values.

In the section "Antenna Characteristic Impedance", I will round the calculated results from three software versions to the same number of digits for comparison.

The antenna models that will be presented using transmission line theory will be approximations, too. The formula for the antenna characteristic impedance is an average value. But, as you will see, it will have good agreement with software approximations.

The simplest model, often called the first approximation, is a good place to begin. Then add adjustments, as needed, for a more precise model. As an example, the resistance of air is usually neglected when calculating gravitational effects on projectiles at low velocities.

In dipoles and monopoles the resistance of the radiator conductor will often be left out of the model when small in comparison to the antenna radiation resistance. For monopoles over real ground, the resistive loss of real ground will need to be included for a more precise model. For a small loop antenna (coil antenna) the Ohmic resistance of the radiator conductor should be included because of that antenna's low radiation resistance.

"Essentially, all models are wrong, but some are useful."
George Edward Pelham Box
1919 - 2013
British Mathematician

1.2 Some Basic Transmission Line Theory

In most cases, efforts are made to keep transmission line reflections low for the purpose of maximum transfer of power between a source and load. When a transmission line is operated with no reflections, it is referred to as a non-resonant line.

When a transmission line is operating with a standing wave it is referred to as a resonant line. Antennas are a type of transmission line operating with a standing wave and have resonances.

An understanding of the 1/2-wave, 1/4-wave, and less than 1/4-wave transmission lines is all that is needed to visualize and calculate monopole and dipole AC circuit values.

First, a line that is exactly 1/4-wave (90 electrical degrees):
1) Is an impedance inverter, sometimes called a transformer. It can invert a high impedance to a lower impedance or a low impedance to a higher impedance. It can invert a capacitance to an inductance, or an inductance to a capacitance.
2) When shorted at one end, the other end looks like a very high impedance, almost an open. A shorted 1/4-wave transmission line behaves like a parallel resonant circuit, which has a high impedance.
3) When open at one end, the other end looks like a very low impedance, almost like a short. An open 1/4-wave transmission line behaves like a series resonant circuit.

Second, regarding a line less than 1/4-wave:
1) A short at one end creates an inductive reactance at the other end.
2) An open at one end creates a capacitive reactance at the other end.
3) The amount of the inductive or capacitive reactance depends on the electrical length, in degrees, and the characteristic impedance of the transmission line.

Third, regarding the 1/2-wave line (180 electrical degrees):
1) A 1/2-wave line "repeats" at its input the impedance at the other end. Think of the 1/2-wave line as two 1/4-wave lines in series, each inverts, and you are back to where you started.

5

Monopoles and dipoles are open ended transmission lines when using the antenna as transmission line model. The high impedance at the open end of an antenna is inverted to a different impedance at the feed point. The loss of energy to space is called a "radiation resistance", Rr.

When a moopole, or half a dipole, is less then 1/4-wave the open end will result in a capacitive reactance at the feed point, as would be true for any open, less then 1/4-wave transmission line. The reactance will be added to the radiation resistance of the antenna to form its total series equivalent impedance of radiation resistance and capacitive reactance (Rr-jX).

The left column of the figure 70, on the next page, applies to the transmission line models of antennas since the distant end of the dipole and monopole are open. The right hand column will apply to a full wave loop example later. For the case of the exactly 1/8-wave transmission line, the reactance is equal to Zo, the characteristic impedance of the transmission line (or antenna).

The less than 1/4-wave lines will have reactance, inductive if shorted, capacitive if open. It is possible to calculate the reactance value if the characteristic impedance, Zo, and the line length in degrees are known using the trigonometric tangent function.

What is not shown in figure 70 is how to calculate the reactance, in ohms, of the open, or shorted, less then 1/4-wave line.

For an open less than 1/4-wave line the capacitive reactance can be calculated if the characteristic impedance, Zo, and the length of the line in electrical degrees are known. An open transmission line capacitive reactance $X = (Zo)/(\tan\Theta)$, where Θ (Theta) is the electrical length of the transmission line in degrees and the capacitive reactance X is in Ohms.

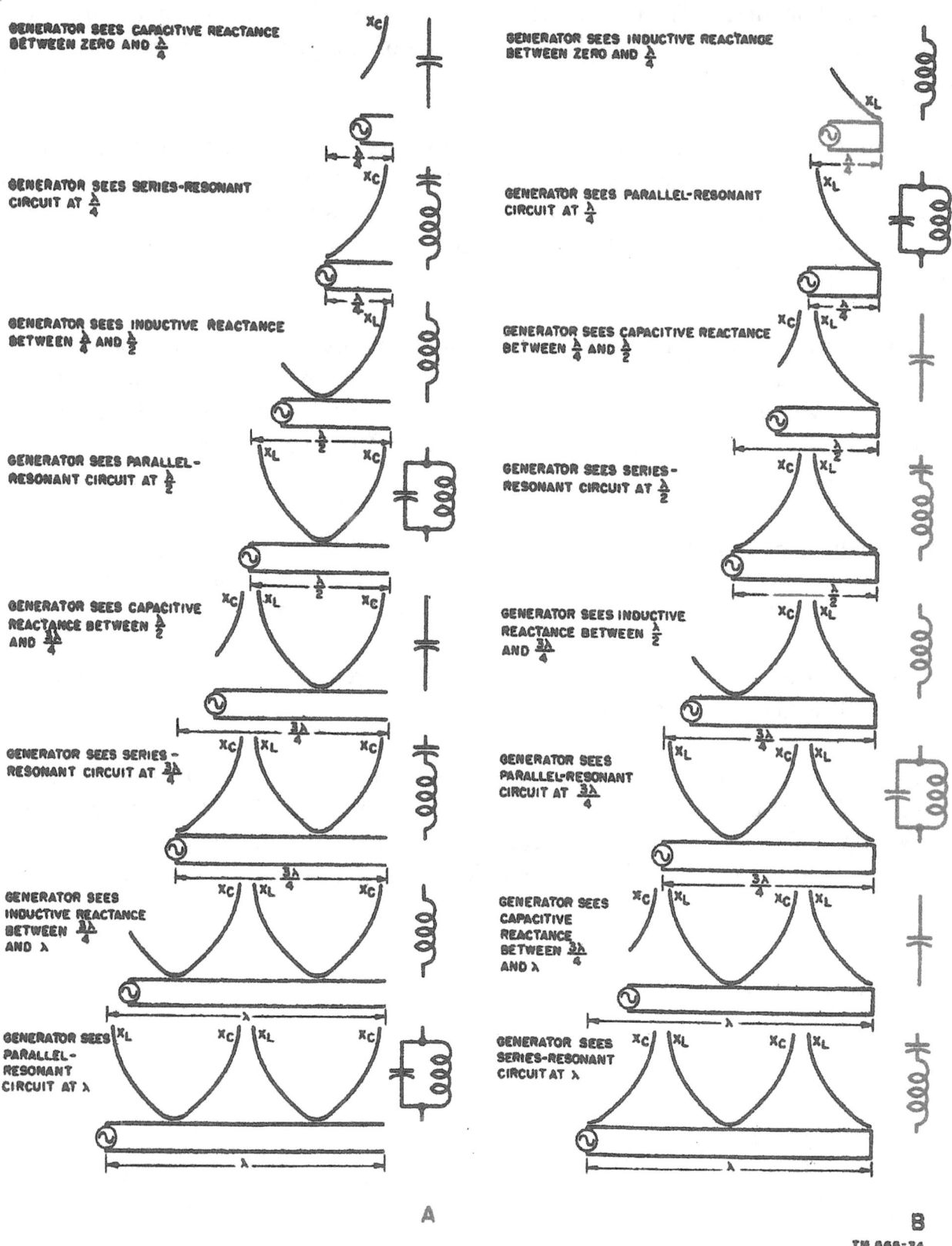

GENERATOR SEES CAPACITIVE REACTANCE BETWEEN ZERO AND $\frac{\lambda}{4}$

GENERATOR SEES SERIES-RESONANT CIRCUIT AT $\frac{\lambda}{4}$

GENERATOR SEES INDUCTIVE REACTANCE BETWEEN $\frac{\lambda}{4}$ AND $\frac{\lambda}{2}$

GENERATOR SEES PARALLEL-RESONANT CIRCUIT AT $\frac{\lambda}{2}$

GENERATOR SEES CAPACITIVE REACTANCE BETWEEN $\frac{\lambda}{2}$ AND $\frac{3\lambda}{4}$

GENERATOR SEES SERIES-RESONANT CIRCUIT AT $\frac{3\lambda}{4}$

GENERATOR SEES INDUCTIVE REACTANCE BETWEEN $\frac{3\lambda}{4}$ AND λ

GENERATOR SEES PARALLEL-RESONANT CIRCUIT AT λ

GENERATOR SEES INDUCTIVE REACTANCE BETWEEN ZERO AND $\frac{\lambda}{4}$

GENERATOR SEES PARALLEL-RESONANT CIRCUIT AT $\frac{\lambda}{4}$

GENERATOR SEES CAPACITIVE REACTANCE BETWEEN $\frac{\lambda}{4}$ AND $\frac{\lambda}{2}$

GENERATOR SEES SERIES-RESONANT CIRCUIT AT $\frac{\lambda}{2}$

GENERATOR SEES INDUCTIVE REACTANCE BETWEEN $\frac{\lambda}{2}$ AND $\frac{3\lambda}{4}$

GENERATOR SEES PARALLEL-RESONANT CIRCUIT AT $\frac{3\lambda}{4}$

GENERATOR SEES CAPACITIVE REACTANCE BETWEEN $\frac{3\lambda}{4}$ AND λ

GENERATOR SEES SERIES-RESONANT CIRCUIT AT λ

A

B

TM 666-74

Figure 70. Changing line length changes the impedance seen by source.

7

As an example, the open 1/8-wave (360 degrees/8 = 45 electrical degrees) 50Ω line, X = (50Ω)/(tan 45 degrees) = -j50Ω. The tangent of 45 degrees is 1, therefore X = -jZo as shown in the figure, and is a capacitive reactance, -j for capacitive reactance.

For the 1/8-wave shorted line, X = (Zo)(tanΘ), and for 45 degrees, the tangent being 1, +jX = Zo, an inductive reactance of +j50Ω, +j for inductive reactance.

The same will apply to our antenna as transmission line models. If the monopole (or half the dipole) is less then 90 degrees, the antenna reactance at the feed point will be capacitive, -jX. The antenna as a less than 90 degree transmission line with an open at the distant end will have a capacitive reactance equal to (Zo)/(tanΘ) at its feed point.

For an open transmission line, reactance is found using X = (Zo)/(tanΘ).
For a shorted transmission line, reactance is found using X = Zo(tanΘ).
Since Zo is in Ohms, the calculated reactance values will also be in Ohms.

Following convention, capacitive reactance will be -jX (negative) and inductive reactance will be +jX (positive).

Figure 70 tells us that at exactly 90 degrees an open end transmission line (monopole or dipole) is like a series resonant circuit. At resonance an antenna will have only radiation resistance at the feed point. This is the condition of an ideal resonant antenna, total X = 0, and only the radiation resistance (Rr) of the antenna remains at the feed point.

1.3 Antenna Characteristic Impedance

The formulas for calculating various configurations of transmission line characteristic impedance values have been around for a very long time. In general, transmission line characteristic impedance is related to the distance between and the diameters of the conductors.

In an antenna as transmission line model the line is spread out. The monopole antenna can be viewed as an opened out coaxial transmission line. The outside shield is spread to form the "ground plane", the inner conductor is the radiator.

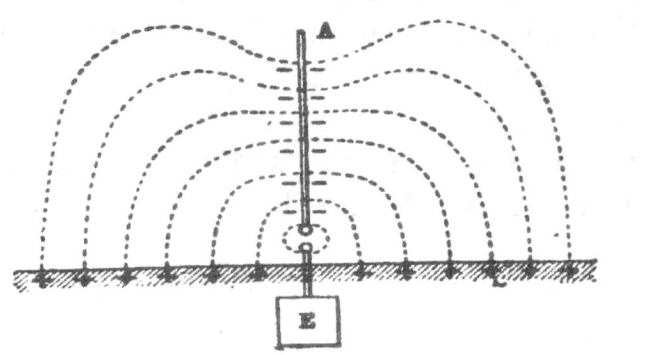

FIG. 8.

Lines of Electric Force of Linear
Antenna.

The monopole figure 8 is a 2-D image. For a complete view you must rotate the image around the axis of the monopole and realize those dashed lines of electric force are also toward you and away into the page. They surround the radiator, 360 degrees around, making the ideal monopole omni-directional. Those E-field dashed lines are at right angle to the conducting surfaces.

Capacitance, as a general idea, is two conductors separated by an insulating medium (dielectric). There are two aspects of capacitance to keep in mind related to antennas.

1) As the distance between the conductors increases, the capacitance decreases.
2) As the surface area of the conductors increase, the capacitance increases.

The distance between the top of the monopole and the ground plane is greater than at the base. As you move up the monopole, from the base, the distributed capacitance of the antenna is decreasing due to increasing distance. The distributed capacitance of the antenna is not constant as with a regular transmission line. This is why the formula for antenna characteristic impedance (provided later) is considered an average value.

From a circuit perspective for a capacitor, the larger the capacitance the more charges it can store. Since the capacitance of the monopole is greatest at the bottom, closer to the ground plane, more charges will be there than at the top where the capacitance is lowest. Maximum charge, and therefore maximum current will be at the base of the monopole antenna.

Voltage will be greatest at the open end because it is a high impedance. The current in the antenna will be lowest at the open end because of the high impedance due to the small capacitance, and the very large reactance (-jX) at alternating currents.

9

The current will be maximum and the voltage minimum at the bottom, the usual feed point of the monopole in this figure. According to Ohm's Law R = V/I, the high current and low voltage at the base, leads to a low radiation resistance (Rr) at the bottom feed point of the monopole antenna.

Here we see the vertical monopole antenna as more of a transmission line model, showing distributed inductance and distributed capacitance. This image shows the capacitive and inductive parts of the monopole.

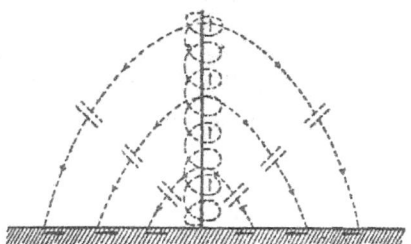

The monopole conductor is one plate and the ground plane is the other plate of a capacitor antenna with inductance due to the conductors. When the reactance of the distributed capacitance is equal to the distributed inductive reactance the antenna will be at its resonant frequency and total reactance will be X = 0, leaving only the radiation resistance, Rr, plus any system resistive losses.

In this book I begin with monopoles. The monopole is the most simple model. I will get to dipoles in due course by showing that a dipole is two monopoles connected in series, feed point to feed point. The monopole is half a dipole, the other half is an "image" that does not radiate. Marconi took Hertz's dipole, using one half of the dipole as a radiator. This is why a monopole antenna is also called a Marconi antenna.

When working with all transmission lines I will be using the natural log, Ln. I do this because the derived formulas for transmission lines, in their original form, use the natural log, not the base 10 log, also called the "common log."

Transmission line formulas you may find that use base 10 common logs are approximations. The original, Ln, natural log formulas have been multiplied by an approximate number to convert from natural logs to base 10 common logs.

If the original formulas use Ln, then why convert them to base 10 common logs? In the days before scientific calculators and computers, the base 10 log formulas for transmission line characteristic impedance were created for use with tables. With modern calculators we can return to the original natural log, Ln.

The *average* characteristic impedance of a monopole antenna over perfect ground, as a transmission line model, can be calculated from the following mathematical relationship: **Za = 60[Ln(2L/d) -1]**, where L is the length and d is the diameter of the monopole antenna conductor. Length and diameter must be in the same units.

Here, and going forward, I designate the characteristic impedance of an antenna as Za to distinguish it from the Zo of regular transmission lines.

The graph of the trigonometric tangent function and its curves are the same as the curves for the reactance of antennas. The tangent function will make possible the calculation of reactance values for non-resonant antennas using the antenna as transmission line model and the antenna characteristic impedance, Za

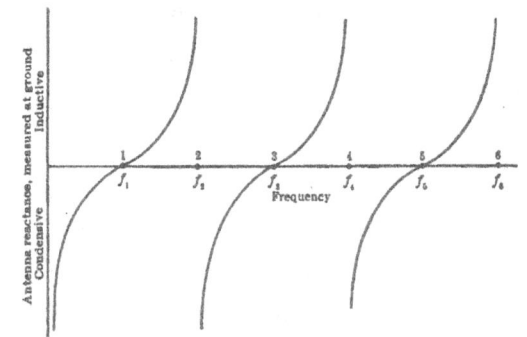

FIG. 60.—As the frequency impressed on an antenna is varied the reactance (as measured at the base) goes through the changes indicated here; in case an antenna with appreciable resistance had been considered the reactance changes from its high positive value to high negative value by going through zero values at 2, 4 and 6.

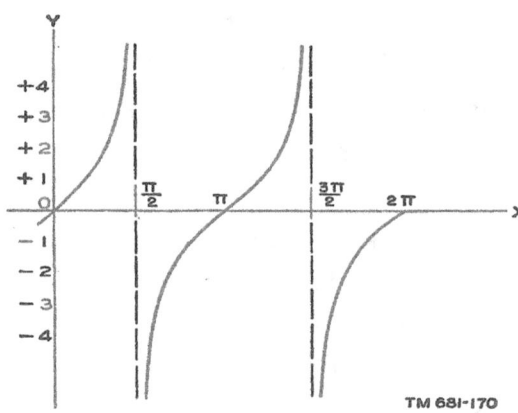

Figure 223. Graph of y = tan x.

The curves of antenna reactance (left) follow the trigonometric functions of transmission lines (right). Notice they repeat. At resonance the reactance becomes zero, the usual condition for series resonance in AC circuits (where the curves cross the horizontal line X = 0).

Monopole Examples of the Antenna as a Transmission Line Model

For a monopole radiator I'll use a 10ft (3.048m) length of electrical conduit, outside diameter, 0.922 inches (23mm measured). I use feet and inches in this example because those are the units used for the conduit I purchased. The units must be the same so I choose to convert 10ft to 120 inches.

The average characteristic impedance, Za, of this conduit monopole antenna as a transmission line will be,
$Za = 60[Ln(2L/d)-1] = 60[Ln(2)(120in)/(0.922in) -1] = 274\Omega$.

The natural resonant, 1/4-wave, frequency is, $f = (234)/(10ft) = 23.4MHz$. At the natural resonant frequency the antenna reactance will be zero. The feed point of the monopole will have only radiation resistance in the ideal model at resonance. If used at any frequency lower than 23.4MHz the antenna will act like any open less than 1/4-wave transmission line and have a capacitive reactance with a radiation resistance.

The capacitive reactance, at a frequency lower than resonance, can be calculated using a basic transmission line relationship for a less than 1/4-wave transmission line, $-jX = (Zo)/(tan\Theta)$, which for the antenna as a transmission line model becomes, **$-jX = (Za)/(tan\Theta)$.**

What will be the feed point capacitive reactance of the 10ft conduit monopole at 21.1MHz?

First we need the length of the conduit in electrical degrees at 21.1MHz. Since 23.4MHz is the 90 degree length, the 21.1MHz length in degrees will be less than 90 degrees. This monopole will have a feed point capacitive reactance at the lower frequency of 21.1MHz. At 21.1MHz, the monopole electrical length in degrees becomes $\Theta = (21.1MHz/24.3MHz)(90\ degrees) = 81.15\ degrees$.

To find the feed point capacitive reactance of the 10ft conduit monopole in Ohms at 21.1MHz, $-jX = (Za)/(tan\Theta) = (274\Omega)/(tan\ 81.15\ degrees)$. The calculated value of the conduit monopole capacitive reactance to 3-places is, $X = 42.7\Omega$, written as $-j42.7\Omega$, since it is a capacitive reactance.

What will be the capacitive reactance of the 10ft conduit monopole at 18.1MHz?

The radiator length has not changed so Za does not change. What does change is the electrical length in degrees because the frequency changed. The lower frequency will not only result in a capacitive reactance, but a larger capacitive reactance than at 21.1MHz.

The electrical length at 18.1MHz will be, $\Theta = (18.1MHz/23.4MHz)(90\ degrees) = 69.62\ degrees$. The calculated capacitive reactance of the monopole at 18.1MHz will be, $-jX = (274\Omega)/(tan\ 69.62\ degrees) = -j102\Omega$.

As the frequency becomes lower than the self resonant frequency the capacitave reactance increases. Since capacitance and capacitive reactance have an inverse relationship, the decrease in capacitance (think smaller plate area) of the e;ectrically shorter antenna results in an increased capacitive reactance at the feed point.

One more time, for 14.1MHz. The length in degrees will be (14.1MHz/23.4MHz)(90 degrees) = 54.23 degrees. The calculated capacitive reactance of the 10ft conduit monopole at 14.1MHz will be, -jX = (274Ω/tan 54.23 degrees) = -j197Ω.

The capacitive reactance values of the 10ft conduit monopole calculated using the antenna as transmission line model are now compared to three different software models, to 3-digits,

	Calculated	#1	#2	#3
21.1MHz	-j42.7Ω	-j45.8Ω	-j47.3Ω	-j46.0Ω
18.1Mhz	-j102Ω	-j108Ω	-j110Ω	-j109Ω
14.1MHz	-j197Ω	-j206Ω	-j209Ω	-j210Ω

Treating an antenna as a transmission line gives useful reactance values with just a calculator and basic transmission line theory.

What about the radiation resistance, Rr?

Radiation resistance for a monopole can be estimated from the relationship **Rr = Ro [(1-cosΘ)²/sin²Θ]**, where Ro for a monopole over perfect ground (ideal) is 36.6Ω.

The estimated values of the monopole radiation resistances, Rr, using the calculated Θ values found previously, as compared to software, again, to 3-digits, are:

	Estimated	#1	#2	#3
23.4Mhz,	36.6Ω	35.9Ω	35.8Ω	36.0Ω
21.1MHz	26.8Ω	25.4Ω	25.5Ω	25.9Ω
18.1MHz,	17.7Ω	16.1Ω	16.2Ω	16.8Ω
14.1Mhz,	9.60Ω	8.47Ω	8.49Ω	8.89Ω

Notice that the three versions of software are not in agreement with each other for Rr.

The antenna as a transmission line model, using simple calculation methods, compares well with the three software model versions.

As the monopole length became less than 90 degrees (resonance) two things happen, the capacitive reactance increases and the radiation resistance gets smaller.

The combination of the antenna feed point resistance and reactance is the antenna input impedance, sometimes called the antenna self-impedance.

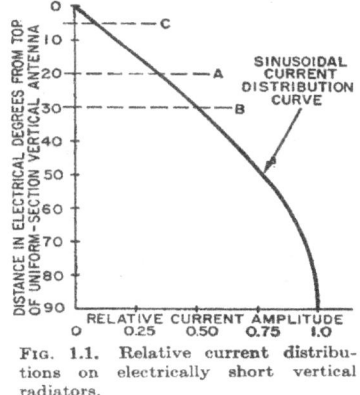

FIG. 1.1. Relative current distributions on electrically short vertical radiators.

As the monopole gets shorter, the current distribution becomes less sinusoidal and more linear. As shown in the figure above, current distribution is sinusoidal to line "B" at around 30 degrees.

Since in all electric circuits the current does the work, the lower 60 degrees of the monopole contributes the most to the radiated field.

Chapter 2 Inductive Loading Short Antennas

In the previous two chapters it was shown that a monopole antenna can be modeled as a transmission line. Through the application of transmission line theory the short linear antenna was characterized as a series AC circuit of radiation resistance and capacitive reactance at the feed point.

In the chapters of this section, monopole antennas over perfect ground will continue to be used as examples. I use the monopole and perfect ground because it is the most simple model and we will build out from monopole examples. A perfect ground is a concept, not something that exists in the real world (only in theory and in software).

When an antenna is short compared to its natural resonant frequency, the feed point impedance can be treated as a resistance and a capacitive reactance in series ($ZA = Rr-jX$ in the figure on the next page). For the ideal case of perfect ground the resistance will only be the radiation resistance with no loss resistance.

When an antenna is resonant, it will be resistive only. At resonance in a series AC circuit the inductive and capacitive reactances are equal and being opposite will cancel.

For real antennas some amount of loss resistance may need to be included in the series circuit model to account for losses in the antenna conductor and/or an imperfect (real) monopole ground.

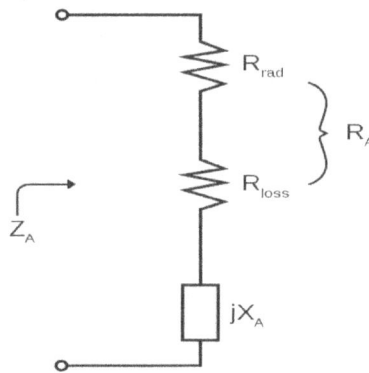

In these next examples the capacitive reactance of a short monopole antenna will be canceled with an added series inductive reactance. By adding an inductive reactance to a short antenna it can be forced to resonance at a lower frequency than the natural resonant frequency.

The most common method of adding inductive reactance to a short antenna to force resonance to a lower frequency is by using a coil of wire. These coils are often referred to as loading coils.

The inductive reactance required for the loading coil will depend on the capacitive reactance of the short antenna and the location of the loading coil along the length of the antenna radiator. The location of the loading coil can make a difference in the required inductance and the final monopole radiation resistance.

A method I have found useful for "adjusting" home brew inductors (image shown on the right) when experimenting. I close wind turns on a form. When the wire is removed from the form it will spring apart (figure on the left). I place zip ties on the coil. The zip ties allow adjustment and hold the coil in position for the required inductance.

A shorted transmission line, of the proper length, can also provide inductive reactance for loading a short antenna. The use of a transmission line to provide inductive reactance has the advantage of lower resistive loss compared to a coil of wire. Using a transmission line for loading an antenna is sometimes called a stub.

Using a shorted parallel wire transmission line to create an inductive reactance load for a monopole will be examined in later examples.

2.1 The Short Antenna and Feed Point Inductive Loading

Short antennas, including the monopole examined in the last chapter, always exhibit two things, capacitive reactance and lowered radiation resistance at the feed point, both dependent on the antenna electrical length in degrees.

The capacitive reactance of the short antenna can be canceled with inductive reactance in series with the antenna at the feed point. The inductor is making up for the missing length of the antenna with a coil of wire.

Using inductive reactance to cancel capacitive reactance is known as conjugate matching. The inductive reactance of the coil cancels the capacitive reactance of a short antenna.

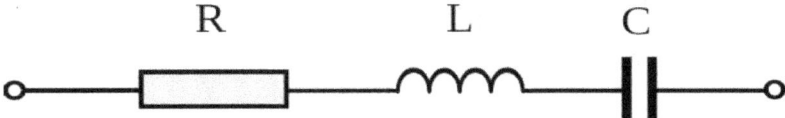

For the case of feed point inductive loading, the reactance of the wire coil needs to be equal in Ohms to the value of the short antenna's capacitive reactance in Ohms. When $XL = XC$, in a series circuit, only the resistance R remains.

Placing a coil of the required reactance in series at the feed point is one simple way to bring the short antenna to resonance. After finding the capacitive reactance of the monopole from the antenna as transmission line model, antenna resonance can be forced to a lower frequency using a coil of equal reactance in series with the monopole antenna at the feed point.

Using algebra, the formula for inductive reactance can be re-arranged to find the needed coil inductance, L, when we know frequency and the capacitive reactance.

The inductance, $L = (jX)/(2\pi f)$, where L will be in micro-Henry (μH) and jX is the capacitive rectance of the short antenna in Ohms to be canceled, f is the frequency in MHz. (The software used for this book will do this calculation for you.)

Using the previous calculations of the 10ft conduit monopole capacitive reactance, from the antenna as a transmission line model, it is possible to calculate the coil inductance required to cancel the antenna capacitive reactance at the feed point for the frequencies lower than 23.4MHz.

The calculated capacitive reactance of the monopole at 21.1MHz was $-j42.7\Omega$. The feed point inductance needed at 21.1MHz will be, $L = (42.7\Omega)/(2\pi)(21.1) = 0.322\mu H$ to force resonance to this lower frequency.

The inductance required for resonance on 18.1 MHz, calculated in the same way, $L = (102\Omega)/(2\pi)(18.1) = 0.897\mu H$, and the inductance for 14.1MHz, $L = (197\Omega)/(2\pi)(14.1) = 2.22\mu H$.

The series addition of the coil at the feed point does not change the radiation resistance of the antenna, it only cancels the capacitive reactance forcing resonance to a lower frequency.

For a purely resistive radiation resistance we can calculate the SWR for a particular feed line characteristic impedance. If, as will likely be the case, 50Ω coax is used, SWR can be calculated using the ideal Rr values previously calculated in the last chapter with the relationship $SWR = (50\Omega)/(Rr)$.

For 21.1MHz, $SWR = 50\Omega/26.8\Omega = 1.87$.
For 18.1MHz, $SWR = 50\Omega/17.7\Omega = 2.82$.
For 14.1MHz, $SWR = 50\Omega/9.60\Omega = 5.21$.

Warning: this SWR calculation method is only true for purely resistive loads! SWR calculation with complex loads, resistance with reactance, is covered in the Appendix.

The coil adds a loss to the antenna due to the resistance of the wire. Feed point inductive loading places the resistance of the coil at the point of maximum current and therefore maximum coil resistance power loss.

Using "high-Q" inductors, inductors with low resistance, is important in antenna inductive loading, especially feed point loading. (more about "Q" next section)

Having a coil at the feed point of a monopole has the advantage of being easily accessible near ground. In the case of our example the inductance could be adjusted or changed for operation on each of the three frequencies. Or the 14.1MHz coil could have taps for 18.1MHz and 21.1MHz, requiring only one coil for three bands.

In the early days of radio, when the wavelengths were long (it was believed then that only long waves could reach distant locations) the radiation resistance of antennas were, in some cases, fractions of an Ohm.

To raise the radiation resistance much was invested in creating capacitance in the antenna by increasing its surface area. The antenna capacitance was brought to resonance on a specific wavelength with an adjustable coil to provide conjugate matching.

The images below show examples of monopoles tried by Marconi in the early 1900s. A single wire (a) was tested, then the circular cage (d) was tried, but destroyed by winds. After that (b), a single fan antenna was temporally used. Figure (e), which I call a square cone, was the final one constructed. Four wooden 210ft towers, forming a 200ft square were erected to support the four upper corners of this aerial.

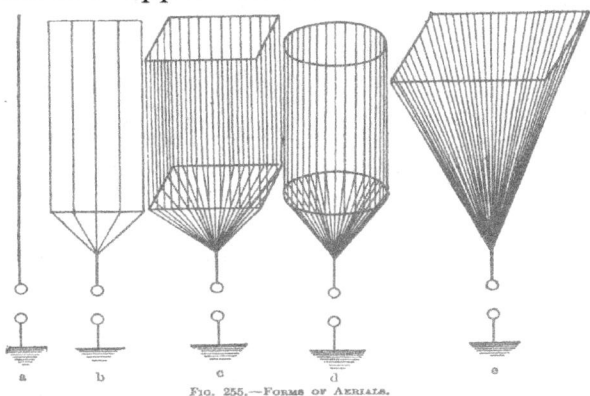

Fig. 255.—Forms of Aerials.

From "Wireless Telegraphy" 1905

The increase in surface area increased the antenna capacitance, and in the circuit model, increasing capacitance, reduces capacitive reactance, requiring less inductive reactance for antenna resonance. This also had the effect of raising the radiation resistance because the equivalent electrical length of the antenna in degrees was increased.

The use of capacitive loading to increase electrical length and raise the radiation resistance of a short antenna will be examined in later sections.

2.2 Antenna Reactance, Resistance, Bandwidth and "Q"

The antenna as transmission line model treats the monopole (and later the dipole) as an equivalent Resistor-Inductor-Capacitor (RLC) series circuit, where R, in an ideal antenna is the radiation resistance, Rr.

Thus far, except for loading coil loss, no other system resistances have been examined in the model, such as resistive loss in the antenna conductor, or any loss in a non-perfect ground. It was mentioned in the previous section that using an inductor at the feed point with a short antenna to force resonance to a lower frequency, that the coil adds a resistive loss.

To reduce the coil resistive loss, "high-Q" inductors, those with low resistance compared to their reactance at the applied frequency, are important in inductive loading, especially feed point loading, where the antenna current is greatest.

The Q of an inductor is equal to the coil reactance divided by the coil resistance, $\mathbf{Q = (X_L)/(R)}$ where R is the AC resistance of the coil wire. From this relationship, as the coil resistance reduces for a given reactance Q increases. Increasing the Q of the inductance increases the Q of the entire series RLC circuit and losses become less. The same will be true if reactance is increased for a particular coil resistance.

What is "Q"? The "Q" of a series resonant circuit is, 2π(maximum stored energy)/ (energy lost) per cycle, where energy lost is due to the coil's resistance.

The more energy stored in the circuit, the larger the Q value for a given system resistance. What "stores energy" in a series AC circuit (or antenna model)? Reactance is the energy storage device in AC circuits. The only work done in an AC circuit is on the resistance in the circuit. Electrons do work on resistive loads, they are stored in reactive loads.

For an antenna the resistances are, radiation and system loss resistances.

Series resonant Q and bandwidth are related by, 1/Q times the series circuit resonant frequency, $\mathbf{BW = (1/Q)(fr)}$. From this relationship we see as the Q increases, the bandwidth becomes smaller. A high Q series RLC circuit is said to have a narrow bandwidth, and is more selective.

Bandwidth in a series resonant RLC circuit is defined as **BW = f2-f1**, the range of frequencies measured between two frequencies, f1 and f2, either side of fr, the resonant frequency. These two frequencies, f1 and f2, are half power frequencies (0.707 points squared = 1/2-power). Since half power is -3dB, these two frequencies are also called the -3dB points of the tuned circuit bandwidth.

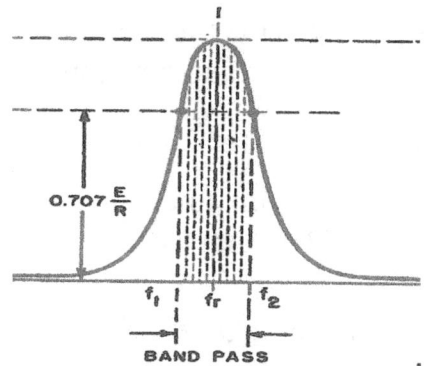
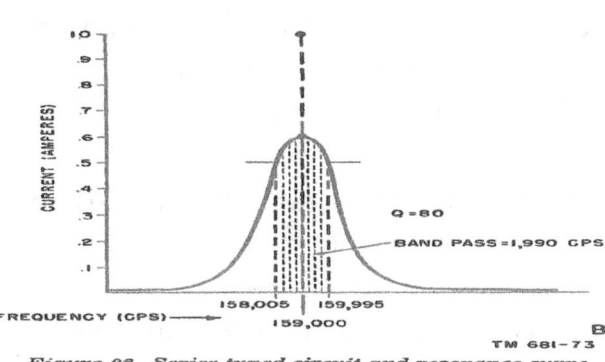

Figure 92. Series tuned circuit and resonance curve.

At the -3dB points half power is delivered to the resistance, the other half is stored in the circuit reactance. For tuned circuits of discrete components, capacitors, inductors and resistors, **Q = (f2-f1)/(fr)** for f1 and f2 at the half-power points, fr being the resonant frequency.

At the series resonant frequency, all the applied power is delivered to the resistive loads because the inductive and capacitive reactances have canceled each other.

Think of an antenna as a window into a range of frequencies within a band of frequencies. A lower Q antenna will have a wider bandwidth, a wider window into the band of frequencies. A higher Q antenna (such as a small loop antenna) makes for a more narrow bandwidth and a smaller window into the band of frequencies.

Let's now focus on the following two relationships, **Q = (jX)/(R)** and **BW = (1/Q)(fr)** for a series RLC circuit, a simple model of our antenna as an AC circuit.

As the reactance, jX, becomes larger more energy is stored, and the Q becomes larger for any value of resistance R. As the value of Q becomes larger, the bandwidth (BW) becomes smaller.

Bandwidth and Q are inversely related. The larger the Q, the more narrow the bandwidth. The lower the Q the wider the bandwidth. What we like from our transmitting antennas is wide bandwidths, which calls for lower Q values. But, we also want low loss in our antennas too.

How can we lower antenna Q to obtain a wider bandwidth? The resistance could be increased. This can be accomplished by raising the radiation resistance of the antenna (which can and will be done in later examples). The total antenna resistance will also be raised by loss resistances. That is not an option we want. The other option is to reduce the reactance of the short linear antenna.

The last option is more easily done because the reactance of the antenna is related to the antenna characteristic impedance. Making Za smaller makes -jX smaller, and antenna bandwidth will be increased.

Let's examine the formula for Za, and see what we can do to lower the antenna characteristic impedance. For a monopole, Za = 60[Ln(2)(Length)/(diameter) -1].

Decreasing antenna length will increase Zo and therefore -jX, increasing Q, which is why electrically short antennas, in general, have inherently narrow bandwidths. But the antenna is often a fixed length for a specific frequency. Changing length changes everything.

Increasing the conductor diameter will decrease Za and -jX. Fat antennas will have a greater bandwidth than thin antennas.

Since SWR is easily measured, often antennas are given an "SWR bandwidth." Some may use the 2:1 SWR frequencies. Others might use 2.5:1 SWR frequencies, another might use the 3:1 SWR frequencies. In effect you get to pick, there is no "standard" SWR bandwidth range.

Beware antennas with wide SWR bandwidths, especially those with physically short lengths or thin conductors. The wide bandwidth may be telling you the antenna has resistive losses wasting energy rather than radiating it.

The most broadband antenna I know of is a resistive dummy load.

We learned that SWR can be simply calculated if the load is purely resistive. For a very thin 1/4-wave monopole over perfect ground Rr is 36.6Ω. With no losses, other than radiation, the SWR with a 50Ω source will be (50Ω)/(36.6Ω) = 1.37 SWR, about 1.4:1.

For a monopole another 1.4:1 SWR possibility is a resistive total of 70Ω, (70Ω)/(50Ω) = 1.4:1, which could happen with a loss resistance of 33.4Ω (70Ω-36.6Ω), almost as much as the monopole radiation resistance.

How is this possible? SWR is a sort of divider or multiplier. An SWR is always the larger divided by the smaller. Therefore (70Ω)/(50Ω) = 1.4:1 SWR, as does (50Ω)/(36.6Ω) = 1.4:1. Low SWR is not necessarily a sign of antenna efficiency.

Always keep the other SWR possibility in mind when thinking about SWR and antenna system losses. An antenna with a wide bandwidth may very well be an antenna with a large loss resistance. And, a 1:1 SWR is not necessarily good in many situations.

A monopole having 1:1 SWR definitely is not good. At 1:1 SWR the loss resistance added to a monopole would be 50Ω-36.6Ω = 13.5Ω, almost 1/3 of the resistance is loss, rather than radiation resistance. The resistive losses (usually ground system loss with a monopole) add to the Rr of the antenna series circuit model changing the SWR.

If a manufacturer specifies a SWR bandwidth of their choosing, should we assume the antenna has 50Ω Rr at resonance? A monopole over perfect ground certainly does not.

What if the antenna at resonance has SWR of 2.5:1 because the Rr = 125Ω or 20Ω (125Ω is possible for a full wave loop antenna, 20Ω is quite possible for a Yagi antenna) and there are no losses? Would the bandwidth be zero for the 2.5:1 SWR bandwidth choice?

Unless the actual -3dB bandwidth of an antenna is measured, the real bandwidth is unknown. I have more to say about antenna bandwidth and SWR bandwidth in Afterwords and Afterthoughts in the Appendix.

The "Q" of an antenna will be calculated in the section on "trap" antennas and used to calculate the needed inductor and capacitor of a tuned circuit to create a two frequency antenna.

2.3 Center vs Feed Point Inductive Loading

A short antenna can be brought to resonance by moving the inductive loading from the feed point to the center of the radiator.

Center loading results in a higher radiation resistance compared to feed point loading.

The amount of inductive reactance required when the loading coil is moved to the center of the radiator can be found using the antenna as transmission line model.

To find base reactance we took a viewpoint looking up from the feed point. What we saw with our virtual eye was an antenna as an open transmission line. Since the antenna was a less then 1/4-wave open transmission line the reactance at the base feed point was capacitive. That capacitive reactance was calculated using basic transmission line theory.

To find the inductance needed for resonance using center loading (or any other point along the antenna conductor) split the radiator and place your virtual eye at the new load location. For center loading, the radiator will be split in half and the transmission line model method will be applied to each half.

Placing our virtual eye at the center of the radiator as a transmission line, when we look toward the open upper half of the antenna we still see an open transmission line less than 1/4-wave long, which will have capacitive reactance.

When we look toward the feed point we see the other half of the antenna as a transmission line terminated in a short on a less than 1/4-wave antenna as transmission line model. (When there is a source, it is removed, and a short is applied. It is standard practice in circuit analysis like this for sources to be removed and replaced with a short.)

What will be the reactance of a shorted less then 1/4-wave transmission line? If you remember, or page back, the reactance will be inductive.

The inductive reactance of a shorted less than 1/4-wave transmission line is found from the relationship: $+jX = (Zo)(\tan\Theta)$, which for an antenna becomes $+jX = (Za)(\tan\Theta)$.

The capcitive reactance of an open less than 1/4-wave transmission line is found from the relationship: $-jX = (Zo)/(\tan\Theta)$, which for an antenna becomes $-jX = (Za)/(\tan\Theta)$.

We will find the capacitive reactance of the open upper transmission line antenna using $(Za)/(\tan\Theta)$, and the inductive reactance of the shorted, lower, transmission line antenna using $(Za)(\tan\Theta)$ for center loading.

This new location, at the center, or any other location between the feed point and the open end, has two parts. We have two transmission lines connected in series providing a total reactance where they meet, in this case the center of the monopole.

Since the capacitive reactance will, by convention, be -jX, and the inductive reactance will be +jX, the total series reactance will be the algebraic sum (keeping the signs) of the two parts where they meet. The total reactance, $X_t = +jX - jX$.

The Z_a of the 10ft (3.048m) length conduit monopole as a transmission line is,
$Z_a = 60[Ln(2) L/d)-1] = 60[Ln(2)(120in)/(0.922in)-1] = 274\Omega$.

The Θ value, since we are in the middle, is the total Θ divided by 2 of the entire conduit electrical length. At 14.1MHz, $\Theta = (14.1MHz/23.4MHz)(90 \text{ degrees}) = 54.23$ degrees. Since we are at the middle of the 10ft conduit each half will be,
$(54.23 \text{ degrees})/(2) = 27.12$ degrees.

Looking toward the upper end from the middle, an open transmission line is seen. For an open line $-jX = (Z_a)/(tan\Theta)$, which will be $(274\Omega)/(tan \ 27.12 \text{ degrees}) = -j535\Omega$ of capacitive reactance for the open upper end half of the antenna.

Looking toward the feed point from the middle, a shorted transmission line is seen. For a shorted line $+jX = (Z_a)(tan\Theta)$, which will be $(274\Omega)(tan \ 27.12 \text{ degrees}) = +j140\Omega$ for the shorted lower half of the antenna.

Since the capacitive and inductive reactances are in series where they meet we add them, keeping their signs in mind. The total reactance at the center will be $X_t = (-j535) + (+j140)$, $X_t = -j395\Omega$. The total is capacitive because the antenna is short at 14.1MHz.

Because the total is a capacitive reactance, a conjugate match inductive reactance of $+j395\Omega$ will be required to force resonance to 14.1MHz when placed at the center of the 10ft conduit monopole antenna.

The required inductance, $L = (X_t)/(2\pi f) = (395\Omega)/(2\pi 14.1) = 4.46\mu H$.

With a 4.46μH inductor at the center of the 10ft conduit monopole, software calculated, $Z = 17.49 + j4.345\Omega$. Moving the loading coil from the feed point to the center of the radiator canceled the capacitive reactance *and* increased the radiation resistance.

To estimate the radiation resistance with a loading coil at other than the base, **$Rr = 36.6\Omega[(sin^2\Theta_1) + (1- sin^2\Theta_2)]$**, where Θ_1 is the length in degrees below the coil, and Θ_2 is 90 degrees minus the length above the coil in degrees.

For the 10ft conduit center loaded monopole, estimated $Rr = 36.6\Omega[(sin^2 \ 27.12) + (1- sin^2 \ 90-27.12)]$, $= 15.2\Omega$ using a calculator, close to 17.49Ω found by software.

Using base inductive loading, calculated previously, for the 10ft conduit monopole on 14.1MHz, $Rr = 9.6\Omega$ estimated by calculator, 8.433Ω calculated by software.

Center loading requires twice the inductance as feed point loading, but almost doubles the radiation resistance. With an increase in radiation resistance, assuming the same antenna resistive losses, the efficiency of the antenna is increased.

With center loading the SWR becomes $(50\Omega)/(17.49\Omega) = 2.9{:}1$, rather than, with base loading, $(50\Omega)/(8.433\Omega) = 5.9{:}1$, according to the software model, assuming perfect ground and zero resistance in the coil and radiator conductor. Calculating these ideal SWR values will tell you right away if the real antenna has added loss should the measured SWR be different.

The center loading inductance being twice the feed point loading inductance may lead you to conclude that placing the coil even closer to the end of the radiator would be a good thing. Keep in mind the distributed capacitance of the monopole gets smaller as we move toward the open end of the antenna. As the capacitance becomes smaller the capacitive reactance toward the open end becomes larger and larger requiring the inductance of the loading coil to be exponentially larger.

As you move from the feed point toward the end of the antenna, the current distribution diminishes (according to the cosine), and coils for loading at low current locations, such as toward the open end of an antenna, need to have very large inductance.

For a real antenna the current at the open end will be small due to the high reactance of the very small capacitance at the top. Loading at or near the open end of the antenna is best done with capacitance, not inductance. Examples of capacitive end loading will be explored later.

If you are doing other than center loading you can split the radiator into any two pieces you want and find the capacitive reactance, the inductive reactance, then the series circuit total, by following the procedure in this section.

We now have a short cut for center loading a monopole radiator. The center loading inductance should be twice that required for feed point loading a monopole of continuous diameter.

If you can measure, or calculate, the capacitive reactance at the feed point of a monopole, a good starting point for a center loading coil is one with an inductance twice that required at the feed point, as long as the radiator is of uniform diameter.

Both the antenna as transmission line model and software treat loads as points, with no real physical size. In the real world the loading inductance will have a physical length.

For real inductors the length of the coil in the antenna adds to the total antenna length. Experiments with RF ammeters have shown that the current above the coil is less than below the coil. Those current differences are consistent with the current distribution of the antenna across the inductor acting as part of the entire radiator length. Expect to do some adjustment, either to the coil or length of the monopole.

What if the top and bottom diameters of a monopole differ? Such a case is examined next.

2.4 Center Loading of Different Diameter Sections

A radiator not being the same diameter either side of the inductive loading coil can, and does, occur. A common case would be a loaded HF mobile antenna with a mast below the coil having a different diameter than the adjustable whip above the coil. Or adding a coil and whip to an existing monopole for resonance on a lower frequency.

The approach is the same as the last example, except the lower section will have a larger diameter than the upper section. There will be two Za values to calculate, one for the larger diameter lower section and one for the smaller diameter upper section.

Let's home brew a monopole for 14.1MHz from some materials on hand. For the bottom section let's use 5 feet (1.524m) of 0.922 inch (23mm measured) diameter conduit below the coil, and 5 feet of 3/8 inch (9.525mm) outside diameter rod for the top.

The antenna characteristic impedance of the conduit lower section will be,
$Za = 60[Ln(2)(5ft)(12in/ft)/(0.922) -1] = 232\Omega$

The antenna characteristic impedance of the upper rod section will be,
$Za = 60[Ln(2)(5ft)(12in/ft)/(0.375in) -1] = 286\Omega$.

Since the two sections have different Za values, I am going to calculate the actual electrical length and not just use half the total electrical length as I did in the last example for halves that were the same diameter. The length of a wave at 14.1MHz is, $(984)/(14.1MHz) = 69.8ft$. The electrical length of 5ft will be, $(5ft/69.8ft)(360\ degrees) = 25.79\ degrees$.

Now to calculate the reactance of each part.

Taking the perspective between the two 5ft parts, we look to the end of the bottom conduit transmission line and see a short. A shorted transmission line less than 1/4-wave will have an inductive reactance which can be calculated with, $+jX = (Za)(\tan\Theta)$. Replacing variables, we have $+jX = (232\Omega)(\tan 25.79 \text{ degrees}) = +j112\Omega$ of inductive reactance for the bottom part.

Looking to the top part we see an open on that antenna as transmission line. An open transmission line, less than 1/4-wave, will have a capacitive reactance, $-jX$. This reactance can be found from $-jX = (Za)/(\tan\Theta)$, which becomes, $-jX = (286\Omega)/(\tan 25.79 \text{ degrees}) = -j592\Omega$.

The total reactance at the center location will be the algebraic sum of $+j112\Omega$ and $-j592\Omega = -j480\Omega$. The $-j480\Omega$ will be canceled with $+j480\Omega$ of inductive reactance where the top and bottom conductors meet. The inductance of the required coil at 14.1MHz to provide $+j480\Omega$ reactance will be, $L = jX/2\pi f = (480\Omega)/(2\pi 14.1) = 5.42\mu H$.

With a 5.42μH inductor at the load location and the two different diameters, software calculated $Z = 17.3+j5.554\Omega$, 2.93:1 SWR.

If you want loading coils for any other frequencies lower than the natural frequency of 23.4MHz, repeat the process for those frequencies, recalculate lengths in degrees and follow the steps to find coils for other frequencies.

Or, use the calculated inductance for those other frequencies to help decide how many turns on the 14.1MHz coil need to be shorted to reduce the inductance on 14.1MHz for the other, higher, frequencies, such as 18.1MHz and 21.1MHz. One coil, multiple taps, multiple frequency choices.

Previously, for the 10ft all conduit monopole, the center loading inductance was calculated to be 4.46μH. In this example, with a thinner rod at the top, the required inductance is 5.42μH. The required inductance increased because the smaller diameter top section had less capacitance (smaller diameter, less surface area, less capacitance) which had a different Za, resulting in a greater -jX.

When the radiator sections are of different diameters the Za of each needs to be calculated. Then the jX of each are calculated using basic transmission line theory to find the total reactance needed at the location where the two radiator sections meet, forming a series AC circuit. The required inductive reactance at the loading location will be the conjugate of the calculated sum from the two transmission line models.

Keep in mind that as the loading coil moves away from maximum current at the feed point toward the end of the monople, the required inductive reactance to force resonance will become exponentially (squared) larger.

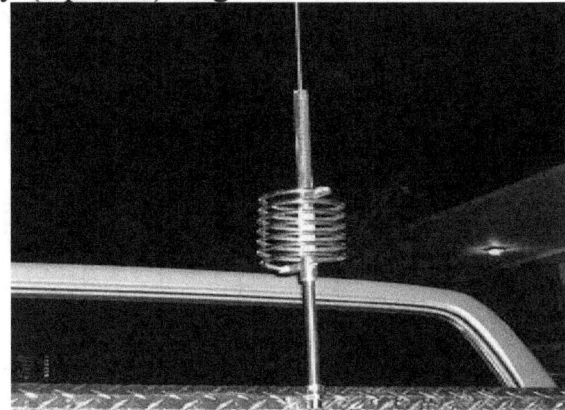

In this example, the "whip" has a smaller diameter than the lower section of the loaded vertical. The builder of this antenna decided not to place the loading coil where the larger diameter lower section meets the thin whip for strength of construction.

2.5 The Shorted Transmission Line Inductive Stub

Previously the 10ft conduit monopole was brought to resonance by using inductive loading in the form of a coil of wire at the feed point.

It is also possible to create an inductive reactance using a shorted less than 1/4-wave transmission line. Obtaining an inductive reactance by using a transmission line will have much less resistive loss than a coil of wire. The Q of a transmission line reactance can be hundreds of times larger than a coil.

To recap:
The Za of the 10ft conduit is 274Ω. At 14.1MHz the calculated feed point reactance was -j197Ω.

In a previous example a coil with inductance to provide +j197Ω was calculated and placed in series at the feed point. Now, instead, a length of shorted parallel wire transmission line will be placed in series at the feed point to provide +j197Ω.

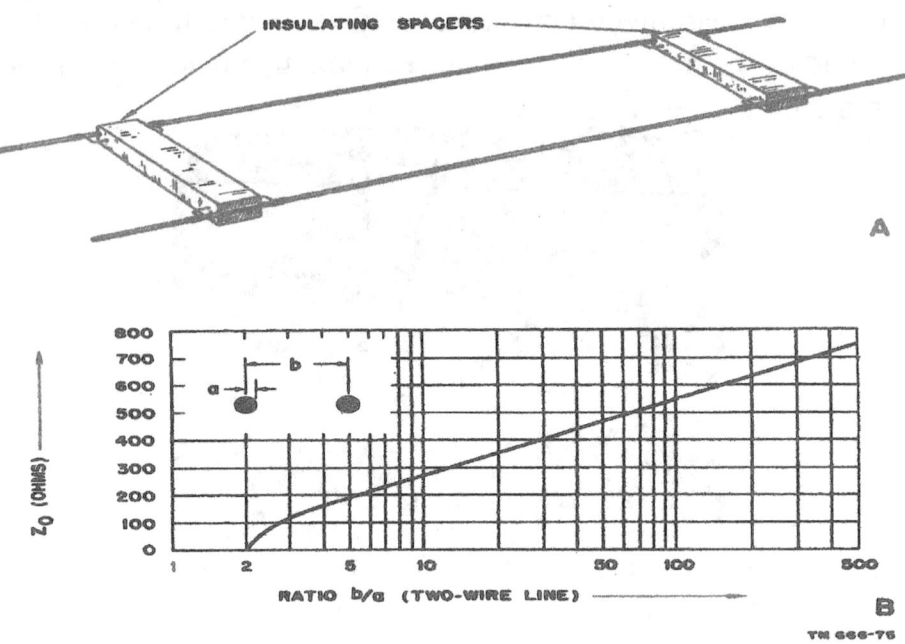

Figure 71. Variation in Z_0 with changes in b/a ratio, open two-wire line.

A shorted parallel transmission line with 6 inch spacing will be used in this example to create an inductive reactance for feed point inductive loading.

The characteristic impedance of a parallel wire transmission line is, **Zo = 120[Ln(2s/d)**, where "s" is the wire spacing and "d" is the wire diameter, in the same units. The Zo of the wire (1.6mm diameter), 6 inch (0.152m) spacing parallel transmission line is 628Ω.

The basic transmission line formula for the inductive reactance of a shorted transmission line is, +jX = (Zo)(tanΘ). Substituting, +j197Ω = (628Ω)(tanΘ), then by Algebraic rearrangement, Θ = tan^{-1} (197Ω/628Ω) = 17.42 degrees.

Now to find the required physical line length from the electrical length, where 69.8ft is the length of a wave at 14.1MHz, (17.42 degrees/360 degrees)(69.8ft) = 3.38 ft (1.03m).

With a 3.38ft shorted wire transmission line stub, 6in spacing, added in series at the feed point, software calculated Z = 8.823-j14.15Ω.

Shorted stubs, as with coil inductive loading at the feed point, have the potential to be easier to adjust. Make them a bit longer than required, move the short to adjust.

In the software model I have extended the parallel wire transmission line out and away in the horizontal plane. You can see why a shorted transmission line inductance is sometimes called a stub.

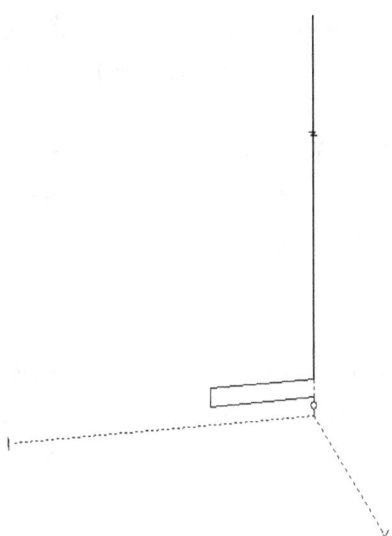

Other than reducing resistive losses, compared to a coil, using the shorted transmission line stub at the feed point to create an inductive reactance has no effect on the antenna radiation resistance, it will still be low.

Use of a shorted stub in this way is also called linear loading.

Chapter 3 - Capacitive Loading Short Antennas

In the last chapter inductive loading was applied to cancel the capacitive reactance of the short monopole forcing it to resonance at a lower frequency.

Why does the short antenna have a capacitive reactance? When the antenna is shortened, capacitance was removed from the antenna. Reduction of capacitance produces a capacitive reactance due to the inverse relationship between capacitance and capacitive reactance.

Rather than cancel the capacitive reactance of the short antenna by adding inductive reactance it is possible to add back capacitance to the antenna. The most effective location for adding capacitance to a short antenna is where the voltage is high, for an antenna this location is at or near the open end.

In the following sections, capacitance will be added to the open end of short monopoles to bring them to resonance at a lower frequency. Capacitve end loading of a monopole has deep roots in the earliest days of radio. In those early days, due to the use of very low frequencies for radio, most antennas were electrically short at long wavelengths.

There was another reason for such low frequencies. Before vacuum tubes, oscillators used for radio were mechanical. In theory, to go higher in frequency you just spin the alternator faster. There was an upper frequency limit due the high speed rotational forces that could cause the alternator to fly apart.

Capacitive loading structures at the top of the monopole were an important way to improve the efficiency of a short antenna. As you will see, capacitive end loading will give a short antenna a higher radiation resistance than either base or center inductive loading.

In this chapter I will treat an antenna at resonance as a quarter wave transmission line impedance transformer. This viewpoint gives us a new and different way to see the connection between inductive feed point loading and capacitive top end loading. In one of the later sections I will show how thinking of an antenna as a quarter wave transformer can solve a "mystery" associated with an inductively loaded dipole.

Several examples in this chapter will show that monopole radiation resistance can be more than 36.6Ω, and a monopole can have multi-frequency possibilities.

3.1 Capacitive End Loading – The "Hat"

As previously mentioned, the replacement of lost length in a short antenna with an inductor has its limits. A point is reached, moving the coil toward the end of the antenna, requiring the coil reactance to become very large. An alternative loading method is to add capacitance back at the open end of the antenna.

There are a variety of ways to add capacity at the end of an antenna. One example of a capacitive end loading structure is a circular disk. Since the disk is placed at the top of the monopole it is common to call it a hat.

The capacitance of a solid disk as a hat in pF is, **C = 0.8992(d)**, where "d" is the disk diameter in inches. To find the diameter of the disk for a required capacitance, d = 1.1121(C), again, d is in inches and C is in pF. Another source gives C = 35.4(d) pF where d in this formula is in meters, or 0.354(d) for d in centimeters. (which converts to 0.9012(d) pF for d in inches).

Using the 10ft conduit monopole as an example, capacitive end loading will be used to force resonance to 14.1MHz with a disk at the open (top) end of the monopole antenna.

To place a capacitive load at the open end of the radiator, our antenna as transmission line perspective is to go to the open end, where the capacitive load is to be placed, and with our virtual eye, look toward the feed point. This was done for the lower half of the center loaded monopole examples.

What we see is a shorted, less than 1/4-wave transmission line which will have an inductive reactance. To find the inductive reactance of this shorted line, the relationship $+jX = (Za)(\tan\Theta)$ is used.

Replacing the variables we get, for this specific 14.1MHz case, $+jX = (274\Omega)(\tan 54.23$ degrees$) = +j380\Omega$. This inductive reactance seen at the top of the monopole will be canceled with a capacitive reactance of $-j380\Omega$ using the capacitance of a solid disk placed at the top end of the monopole.

Another method, using basic transmission line theory, is to consider the missing 6.6ft at 14.1MHz (234/14.1 = 16.6ft), which is (90 degrees - 54.23 degrees) = 35.77 degrees, as having had a capacitive reactance of $-jX = (274\Omega)/(\tan 35.77$ degrees$) = -j380\Omega$ to be replaced by a solid horizontal disk. The $-jX$ of the missing 6.6ft length has the same capacitive reactance magnitude as the $+jX$ calculated by looking down from the open top to the shorted feed point, either way works.

Considering missing length as missing capacitance is a handy way to think about capacitive end loading. The missing length in degrees, calculated as an open transmission line, will be the capacitive reactance that was lost by reducing the antenna length and will need to be replaced by capacitive end loading.

Instead of adding 6.6ft of vertical radiator to the end of the 10ft conduit to achieve resonance at 14.1MHz, a disk will be used to create a capacitive reactanace of $-j380\Omega$.

Since this is an example using a disk type hat we need to find the capacitance in pF for $-j380\Omega$ reactance at 14.1MHz. To convert from capacitive reactance to capacitance the basic capacitive reactance formula is algebraically rearranged into, $\mathbf{C = 1/2\pi f X_c} = 1/(2\pi)$ $(14.1\times10^6)(380\Omega)$. The capacitance needed for a capacitive reactance of $-j380\Omega$ at 14.1MHz is 29.7pF. This means the missing length of 6.6 feet of conduit would have had a total capacitance of 29.7pF.

The required diameter solid disk to replace the missing 29.7pF capacitance, will be $\mathbf{d = 1.1121C(pF)} = 33$ inches.

You may be put off by the idea of a 33 inch (0.838m) diameter disk atop the 10ft conduit to bring the radiator to resonance at 14.1MHz, but that 33 inch diameter horizontal disk is replacing 6.6 ft (79.2in) of vertical height. The reduction in height of a 20M monopole from 16.6ft to 10ft, with the disk hat, might be "stealthy" enough to be less noticeable, or pass as a bird feeder.

Now for the best part of capacitive end loading, the radiation resistance becomes yet higher than with center inductive loading.

The estimated radiation resistance for a capacitively end loaded monopole is, **Rr = (Ro)(sin²Θ)**. For this example Rr will be (36.6Ω)(sin² 54.23 degrees) = 24.1Ω. This is the highest radiation resistance for any of the short, loaded, 10ft conduit monopoles on 14.1MHz examined so far. The resonance SWR, with no other losses, should be around (50Ω)/(24.1Ω) = 2.1:1.

If a solid disk cannot be used a skeletal structure of multiple conducting spokes and a perimeter conductor is often used. Another idea is to use fewer spokes and add screen wire to the spokes and a perimeter conductor.

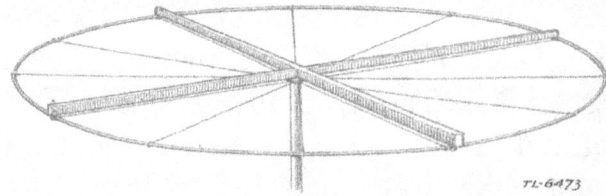

The solid pizza pan hat, on the left, and the frying pan screen, on the right, have the same diameter and, by measurement, added the same electrical length to an experimental conduit vertical antenna. The screen has lighter weight and less wind load.

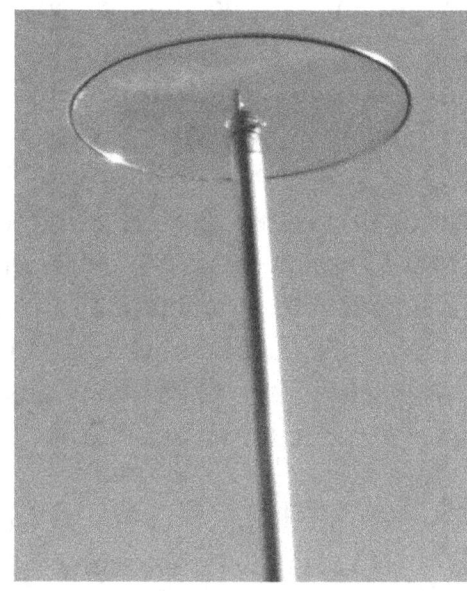

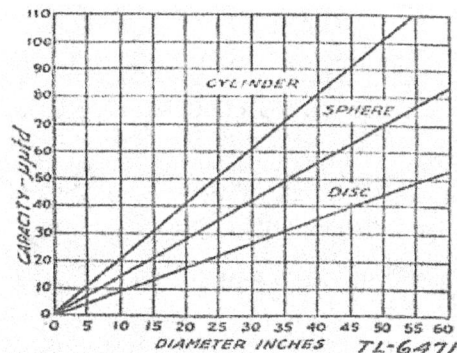

The graph above shows disk, sphere and cylinder capacities for end loading. For a cylinder, height equals diameter.

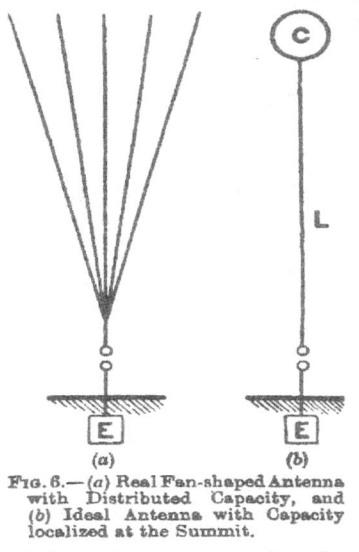

Fig. 6.— (a) Real Fan-shaped Antenna with Distributed Capacity, and (b) Ideal Antenna with Capacity localized at the Summit.

Figure 6, above, from 1908 "The Principles of Electric Wave Telegraphy" shows a fan antenna. The concept of the fan was, unlike the single conductor monopole, for the capacitance of the fan not to diminish with height. Figure (b) shows end loading capacity provided by a sphere. Hertz loaded the ends of a dipole with spheres. Capacitive end loading was essential to the operation of antennas at the long wavelengths in the early days of radio.

Today, with antenna height restrictions, antenna end loading has returned for even the short waves.

The big three of antenna end loading are:
1) The single wire capacitive end loaded antenna, known as the inverted L.
2) The two-wire capacitive end loaded antenna known as the T.
3) And, the one examined in this section, the capacitive geometric form.

The "L" and "T" antennas will be examined in up coming sections.

Note:
The formula for calculating the capacitance of a disk hat will be true for monopoles over perfect ground and, as will be seen later, when disks are used to add capacitance to *both* ends of dipoles.

When disks are used for top end loading of monopoles over other than perfect ground, the actual added electrical length can be about 50 to 65% of the calculated value.

The reduction is due to a non-perfect ground not creating a "mirror image" of the monpole and hat acting as the other half to form a virtual dipole. The closet thing earth has to a perfect ground is sea (salt) water.

3.2 The Pizza Pan Hat - Disk and Inductive End Loading

In the previous example, using basic transmission line theory, a 10 ft conduit monopole was brought to resonance at 14.1MHz using -j380Ω of capacitive end loading provided by a disk as a hat. From the calculated reactance needed at 14.1MHz the capacitance in pF was calculated, and from that capacitance it was found that a 33 inch diameter disk was needed for end loading.

What if a 33 inch disk is not used? What if a smaller diameter disk is used out of necessity (stealth?) or what's available?

In this example I will use a 13.5 inch (0.343m) diameter pizza pan as an example of a reduced size disk capacitive end loading hat.

Reducing the disk diameter reduces its surface area which reduces its capacitance. Reducing capacitance increases capacitive reactance due to their inverse relationship. Using 13.5 inches as the disk diameter and C = 0.8992d, C = 12.1pF for the pizza pan. Capacitive reactance of the 13.5 inch pizza pan at 14.1MHz will be $1/2\pi fC$ = -j933Ω.

Previously it was calculated that -j380Ω of capacitive end loading was required to force a 10ft conduit monopole to resonance at 14.1MHz. The pizza pan reactance is larger than needed.

If you have -j933Ω and need only -j380Ω, what can be done? Add +j553Ω of inductive reactance. In our world of series AC circuit models of antennas, we place an inductance of +j553Ω in series at the end of the antenna and just below the pizza pan.

A coil inductance of L = (553Ω)/(2π)(14.1) = 6.24µH, placed between the end (top) of the monopole antenna and the 13.5 inch pizza pan, will force resonance to 14.1MHz.

The excess capacitive reactance of the smaller disk is canceled by the positive inductive reactance of a coil, basic series AC circuit theory.

Over the years I have used pie pans, cake pans and pizza pans for capacity end loading vertical antennas. A pan that is plated can easily be soldered. Most of my pan capacitive hats have been recycled from thrift stores at very low cost.

There is no one-size fits all when it comes to ready made pan disks. This example shows how to reduce the disk size, add a coil, and capacitively end load an antenna.

With the coil near the end of the radiator, the low current end, the loss in the coil wire resistance is reduced. But, the voltage across the coil will be highest near the end. Higher voltage will place greater stress on the insulation of the coil wire and any coil form. Coils that get warm with application of RF is due to the resistive loss of the wire (and sometimes the coil form insulating material), and is an indication of that loss.

3.3 The Two-Band Pizza Pan Monopole

Previously, a geometric form, specifically a disk, was used to add capacitance to the end of a monopole. The needed diameter of disk was calculated to bring the 10ft conduit to resonance on 14.1MHz. In the last example a smaller disk was placed in series with an inductor at the end of the antenna to force resonance to 14.1MHz.

The 10ft conduit with the 13.5 inch diameter pizza pan hat alone, before adding the inductor below the pan, did not bring the antenna to resonance at 14.1Mhz because the pan was too small in diameter and capacitance.

What is the natural resonant frequency of the 10ft conduit with the smaller pizza pan and without the coil? It must be resonant somewhere between the conduit natural resonance of 23.4MHz and 14.1MHz, the resonant frequency with the larger diameter disk.

At 14.1MHz the conduit alone is, (14.1MHz/23.4MHz)(90 degrees) = 54.23 degrees. It was determined that 29.7pF was needed at the end of the 10ft conduit to replace the missing 35.77 degrees of length (90 degrees - 54.23 degrees) at 14.1Mhz.

If we can find how many electrical degrees are due to the 13.5 inch pizza pan's 12.1pF we can find an equivalent electrical length for the pizza pan. From that electrical length, a new physical length, and the new resonant frequency of the antenna can be calculated.

Let's do a simple ratio operation, (12.1pF/29.7pF)(35.77 degrees) = 14.57 degrees for the smaller pan. Add this equivalent electrical length of the 13.5 inch pan to the electrical length of the 10ft conduit at 14.1MHz for a total equivalent electrical length, (54.23 degrees + 14.57 degrees) = 68.80 degrees.

The physical length will be (68.80 degrees/360 degrees) times the length of the wave at 14.1MHz, which is, (984)/(14.1) = 69.8ft, for a final, total equivalent physical length of, 13.3ft (4.05m).

Natural resonance for a 13.3ft monopole will be, (234)/(13.3ft) = 17.6MHz, very close to the 17M amateur radio band.

Let's now find the feed point series inductance to add 14.1MHz operation to this 17.6MHz monopole with the pizza pan hat.

Back to the antenna as a transmission line model to solve for the required feed point inductance to force resonance to 14.1MHz with a feed point loading coil.

The 10ft conduit has a characteristic impedance of, $Za = 274\Omega$

The electrical length $\Theta = (14.1MHz/17.6MHz)(90 \text{ degrees}) = 72.10 \text{ degrees}$.

The capacitive reactance at the base of the antenna will be, $-jX = (274\Omega)/(\tan 72.10 \text{ degrees}) = -j88.5\Omega$. To resonate the antenna for 14.1MHz, using the smaller pizza pan hat, we need an inductive reactance of $+j88.5\Omega$ in series at the feed point. The required inductance, $L = (88.5\Omega)/(2\pi14.1) = 1.0\mu H$ for operation at 14.1MHz.

Lets now calculate the reactance at the base to load this 17.6MHz monopole for 18.1MHz. In this case the natural resonance of 17.6MHz is lower than the required 18.1MHz. This will produce a feed point inductive reactance, requiring a capacitive reactance at the feed point.

The electrical length in degrees will be $(18.1MHz/17.6MHz)(90 \text{ degrees}) = 92.56$ degrees. To find the inductive reactance of the antenna at 18.1MHz $-jX = (274\Omega)/(\tan 92.56 \text{ degrees}) = +j12.3\Omega$. Or, $+jX = (274\Omega)(\tan 92.56 \text{ degrees} - 90 \text{ degrees}) = +j12.3\Omega$. Either way works. The -90 subtracts the 1/4-wave inverter leaving the actual inductive length in degrees.

The required capacitance for resonance at 18.1MHz, $C = 1/(2\pi)(18.1\text{x}10^6)(12.3\Omega) =$ 715pF.

For a 13.3ft conduit monopole, software calculated, for 18.1MHz, $+j12.65\Omega$, and for 14.1MHz, $-j97.93\Omega$ reactance values, as compared to the transmission line model calculated values of $+j12.3\Omega$ and $-j88.5\Omega$.

This 10ft conduit monopole and 13.5 inch pizza pan hat will be resonant on 14.1MHz with a small series base loading coil. It can also be resonant on 18.1MHz with a series capacitor at the feed point.

Converting the hat into an equivalent length, then adding it to the radiator length to find the new electrical and physical equivalent length can be very useful.

Finding the equivalent length of a hat can be used to create a linear model of the hat, allowing lengthening the antenna model without having to build a multi-wire hat structure in software, saving time.

Another reason for using the antenna as transmission line method to solve capacitive end loading, software does not know what to do with a lumped capacitance added to the *open end* of an antenna.

In a later section "The Monopole – A 1/4-Wave Impedance Transformer" I will introduce a different method to accomplish the results here. That method also uses basic transmission line theory.

In the section "A Two Frequency Antenna Using A "Trap" I will show a way to make the switch from 14.1MHz inductance to 18.1Mhz capacitance automatic through the design of a parallel resonant circuit added to the antenna.

3.4 Capacitive End Loading with One Horizontal Wire
The "Inverted L"

When a geometric form will not do, a horizontal conductor can be used to provide missing end capacitance.

The conductor forms an open transmission line with its image.

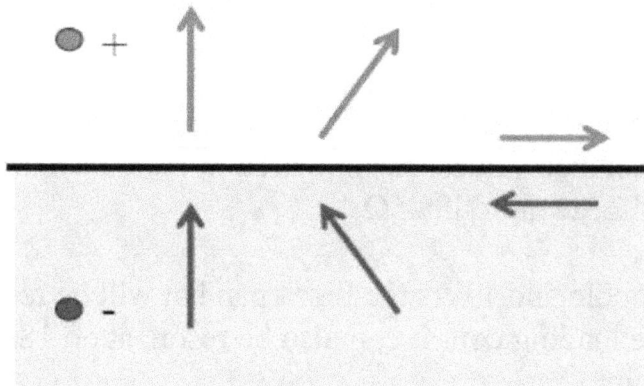

The right most case in this example will form a horizontal single wire transmission line used to provide capacitive end loading for a shorter than 1/4-wave monopole of the inverted L antenna here and the T antenna in the next example.

The image is theoretically the same distance below the surface as the conductor is above the surface. Notice the direction of current flow in the horizontal image. Opposing current flows in the image compared to the wire above ground.

If you have trouble buying into the image theory a simple experiment will help you.

Take a ruler or tape measure and put the end against a mirror. Place a finger at some distance on the ruler or tape measure from the end toward you. You will see the distance of the rule or tape measure is also the same distance as the reflection into the mirror. The total distance is double.

The mirror is acting as a perfect reflecting plane which has already been relied upon for all the previous vertical monopole examples. If you would like to get a visual analogy for a non-perfect ground image, take your rule or tape measure to a window or picture frame glass and repeat the experiment. You can still see an image. The reflection may not be as good as in the mirror but it is there.

The horizontal conductor attached to the top of the monopole forms a parallel transmission line with its image. That horizontal transmission line has an open end and an open end transmission line less than 1/4-wave will have a capacitive reactance.

The characteristic impedance of a single wire transmission line over ground is, **Zo = 60[Ln(4h/d)]**, where "h" is the conductor height above ground and "d" is the conductor diameter, both in the same units. I am sticking with my original plan, transmission line characteristic impedances are Zo, antenna as transmission line characteristic impedance is Za.

This is the first example in which a horizontal conductor has been part of an antenna. Later, we will see that all horizontal conductors used for antennas form a parallel transmission line with a ground image. Horizontal wire height above ground matters in Zo and for the Za of a horizontal antenna.

We already have calculated that the 10ft (3.05m) conduit vertical needs -j380Ω capacitive reactance end loading to force resonance to 14.1MHz.

For this example, a horizontal single wire transmission line will be used to provide the capcaitive reactance, forming an inverted L antenna. At 10ft above ground, the height of the conduit monopole, a wire (0.0641 inch, 1.6mm diameter) will be the single wire horizontal transmission line conductor.

The chaaracteristic impedance, Zo = 60Ln(4)(10ft)(12 in/ft)/(0.0641in) = 535Ω for this single wire transmission line, 10ft above the ground image.

The next question is how long must the horizontal single wire open transmission line be to provide -j380Ω of capacitive reactance? From basic transmission line theory we already know that, -jX = (Zo)/(tanΘ), in this case, Zo is the characteristic impedance of the single wire transmission line used to provide capacitance at the end of the 10ft monopole.

We know we need -j380Ω of capacitive end reactance and Zo = 535Ω. This gives the following by substitution, -j380Ω = (535Ω)/(tanΘ). What we need to find is the tangent value in degrees, which will be the one wire transmission line electrical length.

Applying algebra to -j380Ω = (535Ω)/(tanΘ), becomes tanΘ = (535Ω)/(380Ω). To find the angle Θ the inverse operation of the tangent function is used. For tan^{-1} (535Ω)/(380Ω), the angle Θ = 54.62 degrees. Now that we have an electrical length in degrees we need to find the physical length of the horizontal wire transmission line at 14.1MHz.

To find the open wire line physical length two things will need to be calculated. First, the length of the wave at 14.1MHz, in free space which is, (984)/(14.1) = 69.8ft. Next, find what fraction 54.62 degrees is of 360 degrees (360 being the number of degrees in a wave).

To find the length of the wire transmission line, (54.62 degrees/360 degrees)(69.8 ft) = 10.6 ft (3.23m). An open end, single wire transmission line, 10 ft above ground and 10.6ft (3.23m) long should provide -j380Ω of capacitive end loading to force the 10ft conduit monopole to resonance at 14.1MHz forming an inverted L monopole antenna.

Since capacitive end loading is being used, the estimated radiation resistance is, $Rr = (36.6\Omega)(\sin^2\Theta) = (36.6\Omega)(\sin^2 54.23 \text{ degrees}) = 24.1\Omega$ (same as for the disk). Software calculated, $Za = 27.87+j21.96\Omega$, 2.25:1 SWR for the 10.6 ft top loading wire. With no top loading, $Rr = 8.443-j200.0\Omega$ at 14.1MHz.

A 1/4-wave monopole antenna at 14.1MHz should be, $(234)/(14.1) = 16.6$ feet. The total length of the "L" antenna in this example is 10ft of vertical conduit and 10.6 feet of horizontal wire for a total of 20.6 feet (6.28m), not 16.6 feet (5.06m).

Thinking that an inverted L antenna is just a bent 1/4-wave antenna is obviously wrong. The Zo of the horizontal wire is not the same as the Za of the vertical conduit part of the antenna.

Further, any "rule-of-thumb" values for the capacitance of a horizontal wire in pF per foot or meters you may find in the literature ignore them. Unless those values are based on a wire exactly your diameter and height above ground those approximations will be useless, misleading at least.

We have already calculated that $-j380\Omega$ at 14.1MHz to be about, 29.7pF. For our wire of 10.6 feet, $(29.7\text{pF})/(10.6\text{ft}) = 2.8\text{pF}$ per foot. This will only be true for a bare wire of 0.0641 inch (1.6mm) diameter at 10 feet above ground.

One antenna book gave an approximation of 6pF per foot. Clearly, if that value were used the wire would have been considerably short. Ignore general pF/ft or pF/meter values and do the transmission line calculations.

A cause for variation in the jX component in the construction and use of horizontal wire end loading is the assumption that the wire transmission line will have a velocity factor of one. This will be true only for bare wire in air. In the real world, the wire is likely to have insulation and the velocity of propagation will be less than one. This results in a required physical length less than that calculated for bare wire.

Another real world consideration, the horizontal wire is not very likely to be perfectly straight. Both the antenna as transmission line model and the software model hold horizontal wires perfectly straight. The real world has gravity which will cause a wire supported at distant ends to sag forming a catenary and effect the average height of the single wire and the Zo. As always, expect to adjust for resonance in the real world.

I have kept notebooks of my various antenna experiments over the decades. I had discovered through experiment and construction that the total length of the vertical plus the horizontal part of an inverted L was always longer than that calculated for a linear 1/4-wavelength vertical.

Now, you and I know why.

Unlike a basic, open ended monopole the inverted L will not be omni-directional, the horizontal wire will radiate. This is where the software has its true value, calculating and plotting the radiation field.

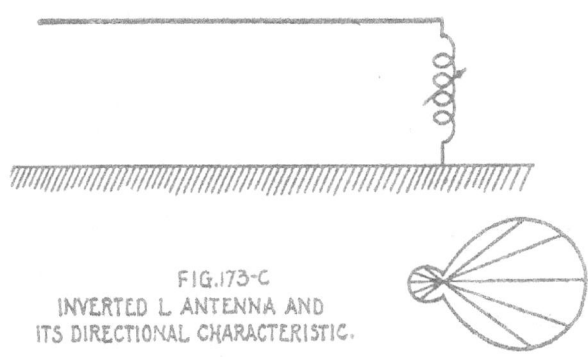

FIG.173-C
INVERTED L ANTENNA AND
ITS DIRECTIONAL CHARACTERISTIC.

An inverted L type antenna was patented by Marconi as a directional antenna in 1905. Due to the long wavelengths used at the time single wire horizontal top load wires were often much longer than the vertical portion.

3.5 Capacitive End Loading with Two Horizontal Wires
The "T" Antenna

In this example two horizontal single wire transmission lines will be used as capacitive end loading with the 10 foot conduit monopole to force resonance to 14.1MHz. The two wires will be co-linear, forming a horizontal line.

The needed capacitive reactance remains -j380Ω for 14.1MHz end loading. The characteristic impedance of the horizontal single wire transmission line, up 10ft (3.05m) remains Zo = 535Ω The length of a wave at 14.1MHz remains, 69.8 feet (21.3m). What will change are the lengths of the two single wire transmission lines to provide a total of -j380Ω end loading.

Since two equal length wires are going to be connected to the top of the radiator, this forms a parallel connection for those two wires at the open end of the conduit. Current flow in and out of the conduit radiator divides equally between the two top loading wires if they are of equal length.

Two equal length wires will be used to balance the end loading wire currents so they do not radiate. Equal length will make the currents equal, but the currents will be opposite in polarity, causing horizontal radiation to be canceled.

The figure on the left show the opposite current division in the conductors of the "T" top loading. The figure on the right shows how top loading increases the average current in the vertical part of the monopole, raising the radiation resistance.

The total capacitive reactance required is -j380Ω. In order for the current to divide in half each top loading wire reactance must double. This means the capacitive reactance of each wire must be twice -j380Ω, -j760Ω.

The calculations are the same as for the inverted L except the capacitive reactance of each wire for the T antenna will be -j760Ω. The electrical length of each of the two single wire transmission lines in degrees will be, $\Theta = \tan^{-1} (535\Omega)/(760\Omega) = 35.14$ degrees. The physical length of each of the two wires in feet will be, (35.14 degrees/360 degrees)(69.8ft) = 6.81 ft.

Using two 6.81 ft (2.08m) wires, forming two single wire co-linear transmission lines at 10ft height, attached to the top of the conduit monopole, software calculated, Z = 25.04+j8.422, 2.07 SWR.

Just as some antenna books claim the inverted L is a bent 1/4-wave, which is wrong, those same books often show the length of the vertical plus the length of the horizontal wire on one side of the T as being 1/4-wave. At 14.1MHz a 1/4-wave monopole is 16.6ft. The 10 ft conduit monopole plus one 6.81 ft wire is longer than a simple 1/4-wave, for the same reason as the inverted L, the Za of the monopole and the Zo of the horizontal wire are not the same.

Unlike the L antenna, the currents in the two co-linear wires of the T divide equally being of equal length. This results in no radiation from the horizontal wires unlike the L antenna. The currents cancel each other because they are out of phase. There will be no pattern distortion, same as a monopole without top loading, or loaded with a disk.

What if the two conductors are not wires? Let's use conduit instead of wire.

Using conduit will give a new Zo for the open horizontal transmission line. The new diameter will result in Zo = 60Ln(4)(10ft)(12in/ft)/(0.922in) = 375Ω.

The new electrical length needed, in degrees, becomes Θ = tan⁻¹(375Ω)/(760Ω) = 26.26 degrees. The physical length in feet needed will be, (26.26 degrees/360 degrees)(69.8 ft) = 5.09 ft, a reduction in length compared to the 6.8ft wires.

From a circuit theory perspective this is reasonable. The larger diameter of the conduit, compared to the wire, has more surface area for charges, creating more capacitance.

Software calculated Z = 25.28 +j9.667 for the two 5.09ft (1.55m) conduits as end loading conductors.

Speaking of using conduit for the end loading, in the days of long wavelengths and short vertical radiators, the top capacitive element, and often the vertical portion of the monopole itself, would be made up of several wires forming a cylinder to increase the surface area for greater capacitive loading. The several wires formed a tube called a cage. Cage type antennas were used for the first short wave (around 100M) transatlantic communications by amateur radio operators in 1923.

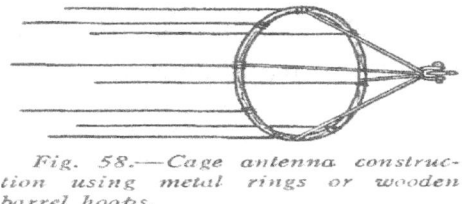

Fig. 58.—Cage antenna construction using metal rings or wooden barrel hoops.

When two conductors were used at the monopole top, this lead to another name for a T antenna, a "flat-top" antenna.

Many amateur radio operators have used a dipole with parallel feeders to get on lower frequencies by connecting the feeder wires of the dipole together at the feed point, creating a T antenna with the dipole arms as the flat-top, and working against ground or a counterpoise. Some books still refer to a horizontal dipole as a flat-top antenna.

The T vertical antenna for LF and VLF were scaled in size for early MF broadcast antennas because the cancellation by the two equal length top loading conductors created no sky-wave, only ground-wave. Thanks to the T antenna end loading current balance its pattern has a deep over-head null like a non-top loaded tower or mast antenna.

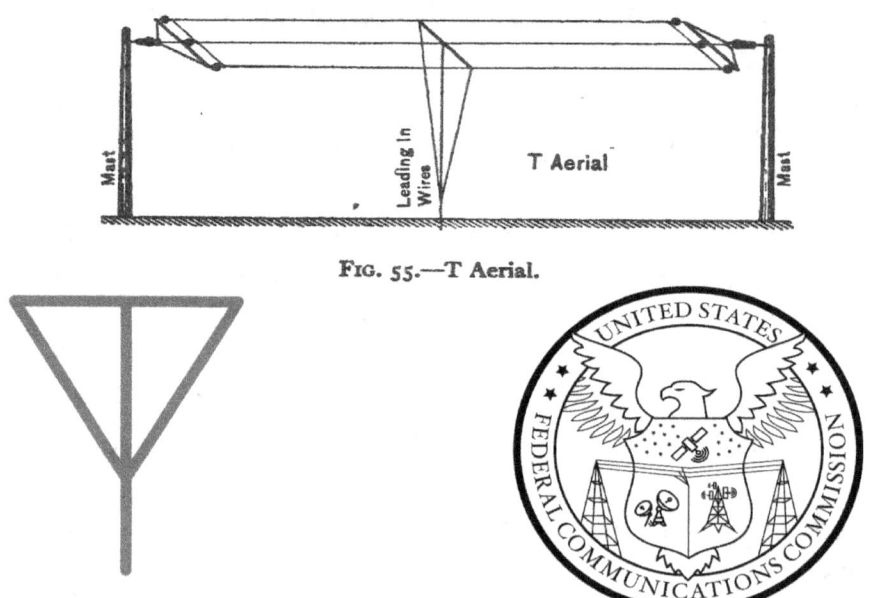

FIG. 55.—T Aerial.

Consider an end-on view of a 3-wire flat-top T with a feed wire from each horizontal wire connected to a single down lead and you will see the schematic symbol for an antenna used today. Do you see the T antenna in the modern day FCC logo?

Later, the T type antenna for MF broadcast gave way to the simpler tower/mast antenna which did not require two supports for the ends of the horizontal top loading wires. The two supports for the top conductors of the T antenna, if metal, were found to distort the monopole pattern and eventually gave way to a single tower/mast as the radiator, which reduced costs. Some single tower/mast MF antennas did, for a time, also use top loading hats.

3.6 Capacitive End Loading Using an Open Transmission Line "Stub"

In this example an all wire, shorter than 1/4-wave monopole, will be capacitively end loaded using an open parallel wire transmission line.

Previously a single wire, and a two wire horizontal transmission line, using its image were used. With that method the height above ground of the horizontal wire was important in calculating the Zo of the end loading transmission line, and its capacitive reactance.

45

In this example, the open wire transmission line used to create capacitive end loading will not be related to the wire's height above ground. The stub will be an open two wire parallel transmission line. The wire open stub at the monopole end will add end capacitance forcing resonance to 21.1MHz.

As we know from previous work with a 10 foot radiator, the natural 1/4-wave resonance is 23.4MHz. The natural 1/4-wave length on 21.1MHz will be, (234)/(21.1) = 11.1ft (3.38m). The 10ft vertical wire antenna is missing 1.1ft (0.335m) for operation on 21.1MHz.

The 10ft wire monopole antenna characteristic impedance will be, $Za = 60[Ln(2)(10ft) (12in/ft)/0.0641in) -1] = 434\Omega$. The electrical length in degrees of the 10ft radiator will be, (21.1MHz)/(23.4MHz)(360 degrees) = 81.15 degrees.

Since we are going to end load this antenna we look down from the top to the shorted feed point. The inductive reactance of a shorted antenna as transmission line can be found from, $+jX = Za(tan\Theta)$. Substituting, $+jX = (434\Omega)(tan\ 81.15\ degrees) = +j2787\Omega$.

Or, find the capacitive reactance of the missing 1.1 feet, which in degrees will be (90 degrees - 81.15 degrees) = 8.85 degrees. The capacitive reactance of the missing length will be $-jX = (434\Omega)/(tan\ 8.85\ degrees) = -j2787\Omega$. Either method works.

To force resonance to 21.1MHz we need a reactance of $-j2787\Omega$ added at the top end of the monopole to replace the missing length of 1.1ft. This time an open wire parallel transmission line will be used to provide the required capacitive reactance.

I will use 6 inch (0.152m) spacing between the wires of the parallel transmission line. Any distance can be selected. The characteristic impedance of a parallel wire transmission line can be calculated from the basic transmission line formula, $Zo = 120Ln(2s/d)$, where "s" is the spacing between the two wires and "d" is the diameter of the wires, using the same units.

For 6 inch spacing, a parallel transmission line using wire, $Zo = 120Ln(2)(6in)/ (0.0641in) = 628\Omega$. The electrical length in degrees of a 6 inch parallel wire transmission line required to provide $-j2787\Omega$ will be, $\Theta = tan^{-1} (628\Omega)/(2787\Omega) = 12.70$ degrees. The physical length of a wave at 21.1MHz will be, (984)/(21.1MHz) = 46.64 ft.

The open stub transmission line physical length will be (12.70 degrees/360 degrees) (46.64 ft) = 1.65 feet (0.503m). With a 1.65ft stub at the top end of the 10ft wire monnopole, software calculated $Za = 35.66+j15.07\Omega$, 1.63 SWR. For this end loaded monopole the estimated $Rr = (36.6\Omega)(sin^2\Theta) = (36.6\Omega)\ (sin^2\ 81.15\ degrees) = 35.7\Omega$.

In the image for the software model you can see how the open transmission line stub makes use of the antenna conductor as one wire of the parallel line.

Two advantages of the stub are that height above ground does not determine the capacitive reactance. and it requires less horizontal space, only the spacing of the parallel transmission line.

A disadvantage of this method of using the antenna conductor as one of the two conductors of the parallel transmission line is that it should not be too long. With this type stub it is best to not go much beyond the top 30 degrees of the radiating portion. The antenna current in that region is fairly linear and helps to ensure the transmission line stub currents are equal and opposite.

3.7 The 1/4-Wave Monopole – An Impedance Transformer

In the section on transmission line theory, the 1/4-wave transmission line was called an inverter or transformer. A line that is 1/4-wave acts as an impedance inverter and will invert a high impedance to a lower impedance or a lower impedance to a higher impedance.

A 1/4-wave transmission line of a particular characteristic impedance can be used to match two different resistances, and when used in this way, is often called a 1/4-wave transformer. A 110Ω resistance (such as the radiatioin resistance of a large loop antenna) can be matched to a 50Ω transmitter or receiver using a 1/4-wave transmission line transformer of a specific characteristic impedance.

The relationship to find the 1/4-wave line transformer characteristic impedance needed is, **Zo = √(R)(R)**. To match 110Ω to 50Ω, a 1/4-wave line transformer would need a characteristic impedance of, Zo = √(110Ω)(50Ω) = 74.2Ω (likely a 75Ω line would be used). The quantity √(R)(R) is called the geometric mean of the two resistances.

The monopole at its natural resonant frequency is 1/4-wave, as an antenna, and has a characteristic impedance. The 1/4-wave matching formula can be used to answer certain antenna questions. Using the 10ft conduit monopole as an example the natural 1/4-wave resonant frequency will be (234)/(10ft) = 23.4MHz. The characteristic impedance Za = 274Ω.

The transformer relationship can give us information about the resistance at the open end of a resonant 1/4-wave 10ft conduit monopole. A little algebra gives $R = (Za)^2/(Rr)$. If the feed point radiation resistance of a 1/4-wave monopole over perfect ground is 36.6Ω, and its Za is 274Ω (conduit), then the resistance at the open end will be, $R = (274\Omega)^2/(36.6\Omega) = 2051\Omega$, 2050$\Omega$ to 3-digits.

A 1/4-wave line can invert reactance too.

If a capacitor (-jX) is placed at the end of a 1/4-wave line it will be inverted to an inductance (+jX) at the other end, the amount determined by the characteristic impedance of the 1/4-wave line and -jX. If an inductance (+jX) is placed at the end of a 1/4-wave line it will be inverted to a capacitance (-jX) at the other end, determined by the characteristic impedance of the 1/4-wave line and +jX.

For reactance transformation the resistance equation becomes, $Zo = \sqrt{(jX)(jX)}$, where Zo = Za, the characteristic impedance of the 1/4-wave monopole antenna as a transmission line. If the required feed point inductive reactance is known, whether through calculation or measurement, the monopole, as a 1/4-wave transformer can be used to find the needed capacitive loading at the other end, the top of the monopole.

In the section "The Short Antenna and Feed Point Inductive Loading" the 10ft conduit monopole, for operation on 14.1MHz, required an inductive reactance of +j197Ω at the base using the antenna as transmission line model.

Treating the 10ft monopole as a 1/4-wave transformer at 23.4MHz, and $-jX = (Zo)^2/(+jX) = (274\Omega)^2/(+j197\Omega) = -j381\Omega$ of capacitive reactance will be required at the other end (top) of the monopole for operation on 14.1MHz. In the section "Capacitive End Loading – The "Hat", it was found that for a 10ft conduit monopole -j380Ω of top end capacitive loading was needed for 14.1MHz by the antenna as transmission line model.

In the section "The Two Band Pizza Pan Monopole" I found the new electrical, then physical length of a 10ft conduit monopole with the 12.1pF pizza pan hat using the capacitance ratio of the two disks. In that calculation I arrived at a new equivalent physical length of 13.3ft and a new resonant frequency of 17.6MHz.

Let's use the 1/4-wave transformer method and see what we get.

First, the 12.1pF of the pizza pan at 23.4MHz, the conduit 1/4-wavelength, will have a capacitive reactance of, $-jX = 1/2\pi(23.4\times10^6)(12.1\times10^{-12}) = -j562\Omega$. This capacitive reactanace on the top will be transformed to an equivalent inductive reactance at the base by the Zo of the antenna, $+jX = (Zo^2)/(-jX) = (274\Omega)^2/(562\Omega) = +j134\Omega$.

The added electrical length of the $+j134\Omega$ at the base can be found from, $\Theta = \tan^{-1}(+jX)/(Za) = \tan^{-1}(134\Omega)/(274\Omega) = 26.06$ degrees. The total new electrical length becomes, (90 degrees + 26.06 degrees) = 116 degrees. The physical length, calculated from this new electrical length, will be (116 degrees/360 degrees)(42.05ft), where 42.05ft is the length of a wave at 23.4MHz, for a total equivalent physical length of 13.6ft.

The monopole new resoanant frequency, with the pizza pan hat, will be (234)/(13.6ft) = 17.2MHz treating the antenna as a 1/4-wave transformer. The result from the assumed linear capacitive hat relationship previously used was, 13.3ft, and 17.6MHz, for comparison.

Thinking of a monopole as a 1/4-wave transformer at its resonant frequency can be very useful since we know how to find antenna characteristic impedance.

When the -jX of any form of capacitive end loading is added to a short monopole, the Za of that antenna, at the 1/4-wave frequency of the antenna, transforms it into an equivalent inductive reactance at the other end, the feed point. And, the 1/4-wave monopole inverts inductive feed point loading to equivalent capacitive top end loading.

As an example, using the 1/4-wave antenna as as an inverter, if I measure the R-jX at the feed point of a short monopole, by using the antenna Za, and the measured -jX at the feed point, I can find the required reactance for capacitive top loading resonance, using $jX = (Zo)^2/(jX)$.

If you don't want to calculate the characteristic impedance of your monopole antenna (or if it might be "messy" due to size variations), a simple way to get a fair approximation of Za for the antenna is to measure the reactance of the antenna at half its self resonant frequency.

At half the resonant frequency the antenna as a transmission line will be an electrical 1/8-wave. The reactance of any 1/8-wave transmission line (whether open or shorted) is equal to the Zo of the line, in this case the Za of the monopole antenna. The measured capacitive reactance at half the resonant frequency will be a good approximation of the antenna's Za for calculation purposes.

Take the measured jX at the feed point for resonance and, using Za, transform it to the jX needed for capacitive top end-loading.

3.8 The Longer Than 1/4-Wave Monopole – A Two-Band Antenna

In previous examples using inductive feed point loading of the short antenna, which canceled the capacitive reactance at the feed point, it left us with a low radiation resistance.

Moving the inductance away from the feed point to center loading almost doubled the radiation resistance as compared to feed point inductive loading. As the coil moves up the radiator toward the open end the needed inductive reactance of the loading coil increases exponentially. Mid-point inductive loading of a short antenna was used as an optimum location. The winner, so far, in raising the radiation resistance of a shorter than 1/4-wave monopole has been capacitive top end loading.

Another way to increase the radiation resistance of a monpole is to go beyond a 1/4-wave (90 degrees) to obtain a radiation resistance greater than 36.6Ω.

The 1/4-wave length of a monopole at 14.1MHz is $(234)/(14.1\text{MHz}) = 16.6$ feet.

Since the conduit I am using comes in 10ft lengths, let's exceed a 1/4-wave at 14.1MHz by using two sections for a 20ft (6.1m) total length monopole. The resonant frequency of this monopole will be $(234)/(20\text{ft}) = 11.7\text{MHz}$. This lower frequency is an indication that 20ft is too long for 14.1MHz.

A longer than 1/4-wave monopole will have a +jX, inductive reactance at its feed point and a higher than 36.6Ω radiation resistance at 14.1MHz. Using simple transmission line logic here is why the reactance at 14.1MHz will be inductive.

If we start at the top open end, then move down 1/4-wave, that 1/4-wave of the antenna as a transmission line will invert (transform) the open at the top to a short on the remaining length. Now we have a short circuit on that last bit of the antenna as a transmission line. A shorted, less than 1/4-wave line will have inductive reactance.

The characteristic impedance of the 20 ft conduit will be,
$Za = 60[\text{Ln}(2)(20\text{ft})(12\text{in/ft})/(0.922\text{in}) -1] = 315\Omega$.

The length of the 20ft conduit, in electrical degrees, at 14.1MHz will be,
$(14.1\text{MHz}/11.7\text{MHz})(90 \text{ degrees}) = 108.5$ degrees. The remaining shorted less than 1/4-wave section $(108.5 \text{ degrees} - 90 \text{ degrees}) = 18.5$ degrees and will produce an inductive reactance at the feed point on 14.1MHz. To calculate the inductive reactance of the shorted 18.5 degree bit of transmission line,
$+jX = (Za)(\tan\Theta) = (315\Omega)(\tan 18.5 \text{ degrees}) = +j105\Omega$.

Here is another approach. The capacitive reactance at the feed point of an open transmission line of 108.5 degrees will be $-jX = (Za)/(\tan\Theta)$,
$-jX = (315\Omega)/(\tan 108.5) = +j105\Omega$. The sign of the tangent of 108.5 degrees on the calculator will be a negative number, indicating the calculated capacitive reactance is inductive, a double negative makes a positive, resulting $+j105\Omega$. Both methods work.

I prefer to think of a longer than 1/4-wave open transmission line by remembering that the 1/4-wave part inverts the open at the end to a short. Then calculate the inductive reactance of the remaining length, after 90 degrees has been subtracted.

For the 20 ft length conduit monopole, over perfect ground at 14.1MHz, software calculated, $Z = 72.18+j108.2\Omega$. The antenna as transmission line model, $+j105\Omega$. Here's what the calculator provides for estimated radiation resistance, $Rr = 36.6\Omega[(1-\cos\Theta)^2/(\sin^2\Theta)]$, which for 108.5 degrees $= 70.6\Omega$. The radiation resistance has exceeded the 36.6Ω of the 1/4-wave 14.1MHz monopole.

The $+j108\Omega$ inductive reactance of the longer antenna can be canceled at the feed point with a series capacitive reactance of $-j108\Omega$ for operation on 14.1MHz over a perfect ground. The needed capacitance will be,
$C = 1/(2\pi f\, Xc) = 1/(2\pi)(14.1 \times 10^6)(108\Omega) = 105pF$.

Using 105pF in series at the feed point, software calculated, $Z = 72.39-j3.09\Omega$, and SWR of 1.45:1. Likely a 100pF capacitor would do it.

If the monopole over perfect ground were a 1/4-wave long at 14.1MHz the SWR would have been $(50\Omega)/(36.6\Omega) = 1.37$, which is lower than $72.39/50 = 1.45:1$ for the lengthened monople. So why bother with the longer, non-resonant monopole?

By making the length of the monopole longer than 1/4-wave three things happen.

First, the radiation resistance is raised. Second, an inductive reactance, rather than a capacitive reactance is produced at the feed point of the antenna and third, the radiation angle is lowered. The first two improve the efficiency of the antenna. A larger Rr makes any resistive system losses a smaller fraction of the total resistance. And, a capacitor will have much lower loss (higher Q) than a coil of wire needed for a less than 1/4-wave monopole when base loaded.

Another possibility presents itself with this 20ft conduit monopole.

Since the self resonant frequency is 11.7MHz, adding an inductance at the feed point can force resonance to 10.12MHz for two band operation.

Treating the antenna as a transmission line the electrical length at 10.12MHz in degrees, $\Theta = (10.12\text{MHz})/(11.7\text{MHz})(90 \text{ degrees}) = 77.85$ degrees. The feed point capacitive reactance of the 20ft monopole on 10.12MHz, $-jX = (315\Omega)/(\tan 77.85 \text{ degrees}) = -j67.8\Omega$ This capacitive reactance at 10.12MHz can be canceled with a series inductive feed point reactance of $+67.8\Omega$, 1.21μH, if a coil is used.

With 1.21μH in series at the feed point, software calculated, $Z = 22.33+j3.533\Omega$, 2.25:1 SWR at 10.12MHz. The estimated radiation resistance, Rr, is 23.9Ω.

The 20ft conduit monopole with a capacitor at the feed point will have a 1.45:1 SWR at 14.1MHz. With an inductor it will have a 2.25:1 SWR at 10.12MHz over perfect ground.

The use of an inductor for operation at a lower frequency or a capacitor for operation at a higher frequency will return in the trap antenna section.

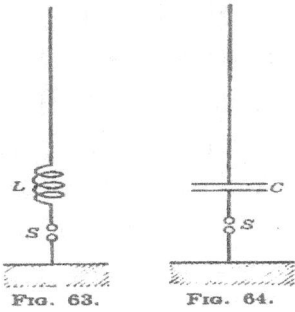

FIG. 63.   FIG. 64.

Here is an example of monopole two-band operation by choice of the series reactance, from "Radio Engineering Principles" 1920. Location "S" is where the output of a spark transmitter would be connected.

The use of a longer then 1/4-wave monopole for multiple band operation will be explored further in the next example.

3.9 A Three-Band Monopole

Previously a 20ft conduit monopole was used as an example of a longer than 1/4-wave monopole on 14.1MHz. The inductive reactance was canceled with a series capacitor at the feed point.

Because the self resonant length of the 20ft monopole was 11.7MHz, it was made 2-band by adding an inductance at the feed point for 10.12MHz.

By careful selection of the monopole electrical length it is possible to produce a 3-band monopole as long as the three frequencies of interest are 1.5 times apart, or less.

Since (21.1MHz)/(1.5) = 14.07, we can have a monopole operate on 14.1MHz, 18.1MHz and 21.1MHz, 3-bands from one monopole, with an SWR of 2.5:1 or less on each of the three bands over perfect ground. The secret is to make the monopole electrical length 117 degrees at the highest frequency. Just take this as a gift for now, this "sweet spot" length will be revealed later.

How long is a 117 degree monopole antenna? Well (117 degrees/90 degrees) = 1.3, and 1.3 times 234 = 304.2. The 117 degree physical length for this example will be, (304)/(21.1MHz) = 14.4ft (4.39m), which has a natural resonance of (234)/(14.4ft) = 16.3MHz.

Treating the antenna as a transmission line using 14.4 ft of conduit, $Za = 60[Ln(2)(14.4ft)(12in/ft)/(0.922in) -1] = 296\Omega$.

On 21.1MHz the antenna is too long by (117 degrees - 90 degrees) = 27 degrees. Using the transmission line equation inductive reactance of a shorted line, $+jX = (296\Omega)(\tan 27$ degrees$) = +j151\Omega$. Software calculated $Z = 106.6+j158.3\Omega$. A capacitive reactance of $-j151\Omega$ will be required at the feed point, which at 21.1MHz is 50pF. Using 50pF in series at the feed point, software calculated, $Z = 106.6-j7.39\Omega$, and SWR of 2.14:1.

At 18.1MHz, the electrical length in degrees will be (18.1MHz/16.3MHz)(90 degrees) = 100 degrees and another inductive reactance at the feed point because it is longer than 90 degrees. The feed point inductive reactance using transmission line theory for 18.1MHz, $\Theta = (100$ degrees - 90 degrees$) = 10$ degrees. $+jX = (296\Omega)(\tan 10$ degrees$) = +j52.2\Omega$, which will require 169pF at the feed point for 18.1MHz. Using 169pF at the feed point, software calculated, $Z = 53.13+j3.087\Omega$, 1.09:1 SWR at 18.1MHz.

Now for the third frequency, 14.1MHz, $\Theta = (14.1$MHz/16.3MHz$)(90$ degrees$) = 77.85$ degrees. The antenna is short at 14.1MHz and will have a capacitive reactance at the feed point. The calculated capacitive reactance is, $-jX = (296\Omega)/(\tan 77.85$ degrees$) = -j63.7\Omega$. Software calculated, $Z = 22.56-j66.09\Omega$ at 14.1MHz.

The inductance required for resonance will be $L = (63.7\Omega)/(2\pi 14.1) = 0.72\mu H$. Using 0.72μH in series at the feed point, software calculated, $Z = 22.56-j2.3\Omega$, 2.22:1 SWR at 14.1MHz.

There you have it. A 14.4ft (4.39m), conduit, 3-band monopole, over perfect ground, using a series inductor for 14.1MHz, series capacitors for 18.1MHz and 21.1MHz, and as promised, all with an SWR less than 2.5:1 on each band.

To find the range of frequencies for this less than 2.5:1 SWR multi-band monopole, take the highest frequency and divide by 1.5. Or, alternatively, multiply the lowest frequency by 1.5.

Another example, (28.1MHz)/(1.5) = 18.7MHz. It will be possible to cover 28.1MHz, 24.9MHz and 21.1MHz for another tri-band monopole, if you make the physical length 117 degrees at the highest frequency of 28.1MHz.

It's your turn, follow the steps, do the calculations, and find the required reactance values for a 28.1, 24.9 and 21.1MHz 3-band monopole over perfect ground.

3.10 A 50 Ohm Radiation Resistance Monopole

An examination of the results of the previous 3-band monopole shows that at, or near, 18.1MHz, the 14.4ft (4.39m) monopole had a radiation resistance of about 50Ω, a typical impedance for transmitters and receivers. Let's use this information from that 3-band monopole at 18.1MHz to find the length of a monopole in electrical degrees for about 50Ω feed point radiation resistance.

The physical length of the monopole in that last example was 14.4ft. The 1/4-wave length at 18.1MHz, (234)/(18.1MHz) = 12.9ft (3.93m). The electrical length of the 18.1MHz monopole is (14.4ft/12.9ft)(90 degrees) = 100.5 degrees, for an approximately 50Ω radiation resistance monopole.

Testing the 100.5 degree electrical length in the radiation resistance estimation formula, $Rr = 36.6\Omega[(1-cos\Theta)^2/sin^2\Theta)] = 52.9\Omega$.

We can now develop a formula to calculate the nominal 50Ω radiation resistance length for any monopole as (14.4ft/12.9ft)(234) = 261. To calculate the physical length of an approximately 50Ω radiation resistance monopole, 261/f = length in feet.

Now for a test run at 21.1MHz, (261)/(21.1MHz) = 12.37 ft, I'll call it 12.4 ft (3.78m).

Here are the calculations by the antenna as transmission line method. The 12.4ft conduit antenna characteristic impedance, $Za = 60[Ln(2)(12.4ft)(12in/ft)/(.922) -1] = 287\Omega$. The inductive reactance at the feed point will be, $+jX = (287\Omega)(tan\ 100.5\ degrees\ -90\ degrees) = (287\Omega)(tan\ 10.5degrees) = +j53.2\Omega$. Using 12.4ft (3.78m) as the length for a conduit monopole, software calculated, over perfect ground, Z = 53.8+j54.82Ω.

Because there is an inductive reactance at the feed point, a capacitive reactance of -j53.2Ω in series at the feed point will cancel the +jX of the 100.5 degree monopole. The capacitance required for a reactance of -j53.2Ω at 21.1MHz will be, C = 142pF.

Using 142pF of capacitance in series at the feed point to perfect ground, software calculated, Z = 54.42+j3.202Ω, 1.11 SWR, which compares well with Rr = 52.9Ω from the estimation formula.

The radiation resistance and reactance of a monopole vary by different trigonometric rules. The radiation resistance varies according to sine and cosine functions. The reactance varies by tangent and cotangent functions.

The graph below shows how the radiation resistance and the reactance of a monopole antenna varies with length.

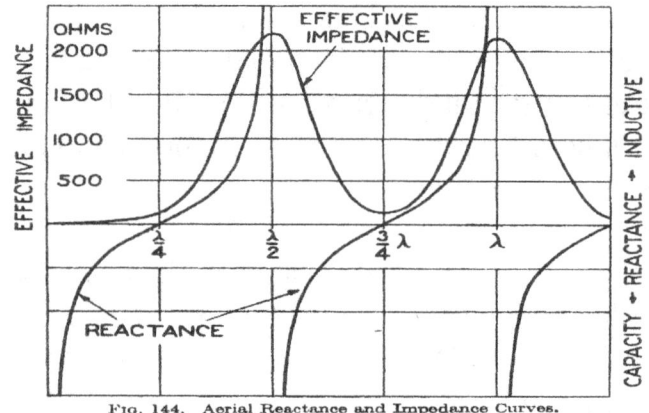

Fig. 144. Aerial Reactance and Impedance Curves.

Looking at the graph, around 1/4-wave, you see the "sweet spot" area where the two functions are fairly close and fairly linear. This is the area taken advantage of for the 3-band monopole.

Note at 1/4-wave and odd multiples of 1/4-wave locations the reactance curves cross zero, indicating resonance. The simple RLC circuit model of an antenna as a series AC circuit is not strictly true. The RLC series circuit has only *one* resonant frequency, antennas can have several. That the reactance of an antenna followed a repeating trigonometric tangent function caused a change in the antenna model from a series RLC resonant circuit model to that of a transmission line model, the basis of this book.

Operation of an antenna on its odd multiples (i.e. 3/4-wave) will be examined later in "The Harmonically Operated Dipole" and the "Maximum Gain Dipole"sections.

At the 1/2-wave locations the radiation resistance becomes largest, typical of a parallel resonant AC circuit.

In the section "The 1/4-Wave Monopole – An Impedance Transformer" the resistance at the open end of the 1/4-wave conduit monopole was calculated to be 2050Ω, compare that calculated value to the graph at the 1/2-wave "effective impedance" peak.

It is always good practice, with any antenna, to assume high voltage at an open end of any radiating conductor. Keep it out of reach, if you can, and to reduce loss use good quality insulators.

Chapter 4 Because There is No Perfect Ground, Use a Counterpoise

Having developed monopoles longer than 1/4-wave, with an inductive reactance at the feed point, including a 50Ω monopole in the example of the previous section, we now can remove the monopole from the theoretical, impossible, perfect ground.

Freeing the monopole from ground gives us an opportunity to apply capacitive loading to the bottom (feed point) end of the monopole. Capacitance added at the bottom of a monopole has been called a capacitive ground, also a counterpoise.

Examples will be shown of how to capacitive load both ends of a monopole later in this chapter. This begins the evolution of the bottom and top end capacitively loaded monopole into the capacitively end loaded dipole.

Historically, the dipole was first. When Hertz did his important experiments he used capacitively end loaded dipoles. This is why a dipole antenna has been called a Hertz antenna.

What Marconi did was to take the Hertz dipole and replace half the dipole with the earth (and lots of buried metal plates). In this section I undo what Marconi did, freeing the monopole from its earthly tether.

I hope to show, and prove, how, at higher frequencies (shorter wavelengths) used today, HF antenna lengths are such that it is not necessary for the Marconi vertical antenna to use ground directly as part of the antenna. Even in Marconi's time it was understood that earth was a poor conductor of electricity. By freeing the monopole from real earth system losses can be reduced improving the efficiency of the antenna.

In the 1920's many of the Amateur radio operators who spanned the Atlantic used antennas with a capacitive ground.

Releasing the monopole from earth, as I will show, lets the monopole be flipped such that the feed point is up, further from earth losses. This inversion of the monopole further improves efficiency.

4.1 Using Capacitive Bottom Loading – The Counterpoise

Geometric forms and horizontal wires as transmission lines have previously been used to create capacitance at the upper end of a monopole over perfect ground. Now that we have developed the 50Ω radiation resistance monopole, which has an inductive reactance at the feed point, we can do capacitive bottom loading using horizontal conductors, or geometric forms, to replace the physical capacitor and the impossible perfect ground.

Using elevated wires or screens to form a capacitance at the feed point of a monopole removes it from direct earth ground and has been called a capacity ground, also a counterpoise, in antenna history.

I have shown how to use single and two wire capacitive loading at the top of monopoles. Now I will apply that same idea to the feed point (bottom) of the 50Ω monopole using two single-wire transmission lines to create capacitance with a balanced counterpoise.

Previously a 2-wire capacitance was placed at the top of a monopole to form a T antenna. In this example the two wires will be placed at the feed point to form an "Inverted T" monopole antenna. As with top capacitive loading, we want the two co-linear wires to have equal length so their currents will be equal, causing field cancellation of radiation from the horizontal counterpoise wires.

For this two wire inverted T example I will raise the feed point of the monopole and the horizontal counterpoise wires 1ft (0.31m) above ground. I will use wire to create two single wire transmission lines to provide capacitive reactance at the feed point to cancel the 50Ω monopole inductive reactance.

Since the two counterpoise wires are connected to the base, the feed point of the monopole, the capacitive reactance of the two wires will make a parallel connection at the bottom, as they did for top loading. To have a total capacitive reactance of -j53.2Ω, each wire will need to provide twice the needed reactance, -j106Ω.

The characteristic impedance of a single wire transmission line 12 inches (0.31m) above ground is, $Zo = 60Ln(4)(h)/d = 60Ln(4)(12in)/(0.0641in) = 397Ω$.

We require a length of single wire open transmission line, Zo = 397Ω, to provide -j106Ω from each wire. The basic transmission line formula for an open end transmission line is, -jX = (Zo)/(tanΘ), which becomes Θ = tan⁻¹ (Zo)/(jX) = tan⁻¹ (397Ω/106Ω) = 75.05 degrees. The length of the wave at 21.1MHz will be, (984/18.1MHz) = 54.37 feet. The counterpoise wire physical length will be, (75.05 degrees/360 degrees)(54.37 feet) = 11.3ft (3.44m) each.

For two, co-linear wires, each 11.3ft long, attached at the feed point, parallel with and 1ft above ground, software calculated, Z = 49.73-j3.673Ω, 1.08:1 SWR. The software shows no field distortion, and looks like a monopole over perfect ground. (See Afterwords and Afterthoughts in the Appendix about counterpoise wire bending.)

If the diameter of the counterpoise conductors were larger, the greater surface area will give the co-linear counterpoise conductors more capacitance. The larger diameter conductors would reduce the Zo of the counterpoise transmission lines. The lower Zo will lead to shorter counterpoise conductor lengths and wider bandwidth.

Using conduit for the counterpoise conductors, Zo = 60Ln(4)(12in/ft)/(0.922in) = 237Ω. The electrical length will be, Θ = tan⁻¹ (237Ω/106Ω) = 65.90 degrees. The physical length will be, (65.9 degrees/360 degrees)(54.37 ft) = 9.95ft (3.03m), a 12% reduction in length due to the counterpoise conductor increased diameter creating more capacitance.

For two 9.95ft co-linear conduits as counterpoise conductors, software calculated, Z = 49.2-j1.596, 1.04:1 SWR. In the near field of the antenna, the current of the antenna flows, mainly, in the counterpoise conductors, and less in earth, reducing ground loss.

This photo (by WB2QCJ) is an example of a wire monopole using two 4ft long rods as a capacitive ground for a 20M, 3/8 wave, monopole. A 3/8 wave antenna has an inductive input reactance, and a 200Ω radiation resistance. The horizontal rods provide a capacitive ground and the 4:1 current balun transforms 200Ω to 50Ω.

Using bottom capacitance has freed the monople from its infinite ground plane requirement, creating new antenna possibilities.

Sir Oliver Lodge, one of the early radio scientists, was not a fan of Marconi's use of earth as the other plate of a capacitor aerial. His position was, "a balanced capacity network of wires or plates raised a foot or so above the earth and insulated from it" should be used.

4.2 A Capacitive Bottom Loaded Three-Band Monopole

We have now removed the monopole from the impossible perfect ground by use of the capacity ground also known as a counterpoise.

I am going back to the "A 50 Ohm Radiation Resistance Monopole" section and apply a counterpoise to that monopole for operation on 20M, 17M and 15M. In that chapter the monopole for 18.1MHz was 14.4ft (4.30m) for a 50 Ohm feed point resistance. Because the monopole was longer than 1/4-wave at 18.1MHz it also had an inductive reactance at the feed point which was canceled by a physical capacitor using perfect ground.

This time the inductive reactance of the 14.4ft conduit monopole on 18.1MHz will be canceled by a 2-wire, co-linear, capacitive counterpoise as done in the previous section.

Here is some essential information for the present project. The electrical length of the conduit monopole at 18.1MHz was 100.5 degrees. The characteristic impedance of the 14.4ft conduit monopole $Za = 60[Ln((2)(14.4ft)(12in/ft)/(0.922in)-1] = 296\Omega$.

The inductive reactance at the base of that conduit monopole at 18.1MHz was $+jX = (296\Omega)(\tan 100.5 - 90 \text{ degrees}) = +j54.9\Omega$. This inductive reactance will be canceled by the capacitve reactance of two co-linear wires, parallel with, but above ground.

The characteristic impedance of the horizontal wire transmission line, 1ft above ground, to be used for bottom capacitance is $Zo = 60Ln(4)(12)/(0.0641) = 397\Omega$. Since each wire should be the same length for radiation cancellation, each single wire transmission line will need a capacitive reactance of $2(-j54.9\Omega) = -j110\Omega$.

For any open transmission line $-jX = (Zo)/(\tan\Theta)$, the needed electrical length is $\Theta = \tan^{-1}(Zo)/(-jX) = \tan^{-1}(397\Omega/110\Omega) = 74.51$ degrees. The physical length of the wire, for a wavelength of $(984/18.1MHz) = 54.36ft$ will be $(74.51 \text{ degrees}/360 \text{ degrees})(54.36ft) = 11.25ft (3.43m)$.

For the lengths found using the transmission line model, software calculated, $Z = 47.15-j12.12\Omega$, 1.29:1 SWR at 18.1MHz.

The 1/4-wave, self resonant frequency of the monopole is, $(234)/(14.4ft) = 16.25MHz$. At frequencies above 16.25MHz the monopole will have an inductive reactance at its feed point, because it will be longer than 1/4-wave. At frequencies below 16.25MHz the monopole will be less than 1/4-wave and have a capacitive reactance at the feed point.

It is quite possible to add series inductive loading to the bottom feed point of the counterpoise capacitive loaded 18.1MHz monopole to operate on 14.1MHz, as has been done with the perfect ground example in a previous section.

To accomplish this, recalculation of certain variables for 14.1MHz will be necessary. Two constants in this recalculation are the characteristic impedance of the antenna, $Za = 296\Omega$, because the length of the conduit has not changed. And $Zo = 397\Omega$ characteristic impedance of the single wire transmission line counterpoise, because its diameter and height above ground will remain the same.

There will be two capacitive reactances at the feed point in series at 14.1MHz, one due to the short monopole itself, and the capacitive reactance of the counterpoise wires at 14.1MHz, which will provide another capacitance in series at the feed point.

The length of the 14.4ft monopole at 14.1MHz, in electrical degrees will be $(14.1MHz/16.25MHz)(90degrees) = 78.09$ degrees. This results in a capacitive reactance at the feed point of the monopole on 14.1MHz of $-jX = (296\Omega)/(\tan 78.09 \text{ degrees}) = -j62.4\Omega$, over perfect ground.

The electrical length of the two 11.25ft (3.43m) single wire transmission lines at the new frequency of 14.1MHz for $Zo = 397\Omega$, where a wavelength is now $(984)/(14.1MHz) = 69.79ft$, will be $(11.25ft/69.79ft)(360 \text{ degrees}) = 58.03$ degrees. Using this electrical length, the capacitive reactance of one counterpoise wire is, $-jX = (Zo)/(\tan\Theta) = (397\Omega/\tan 59.03 \text{ degrees}) = -j247\Omega$ at 14.1MHz. Since there are two of these capacitive transmission line wires connected in parallel at the feed point, the total capacitive reactance of the counterpoise wires at 14.1MHz will be $(-j247\Omega)/(2) = -j124\Omega$.

The $-j124\Omega$ of counterpoise wires reactance is added to the monopole $-j62.4\Omega$ feed point reactance at 14.1MHz to create a total base capacitive reactance of $-j186\Omega$.

Using the previous formula for this 78.09 degree monopole on 14.1MHz the estimated $Rr = 24\Omega$, for a total impedance of $Z = 24-j186\Omega$. Software calculated $Z = 21.47-j191\Omega$.

The $-j186\Omega$ total capacitive reactance at the feed point on 14.1MHz can be canceled with an inductance of $L = (-jX)/(2\pi)(14.1) = (186\Omega)/(2\pi)(14.1) = 2.1\mu H$ in series at the feed point. With an inductance of $2.1\mu H$ added in series at the feed point of the 14.4ft monopole, and the 2-wire counterpoise for 18.1MHz, software calculated $Z = 21.47-j4.425$, 2.35:1 SWR at 14.1MHz.

Now for the antenna at 21.1MHz.

As was the case at 18.1MHz, the monople at 21.1MHz will have an inductive reactance at the feed point. The two counterpoise wires for 18.1MHz will still have capacitance at 21.1MHz, but not enough to cancel the higher antenna inductive reactance at the new frequency of 21.1MHz. This will result in a total inductive reactance and the need for a physical capacitor at the feed point to resonate the antenna at 21.1MHz.

The electrical length of the monopole at 21.1MHz will be $(21.1MHz)/(16.25MHz)$ (90degrees) = 116.9 degrees. The calculated inductive reactance of the monopole at the feed point will be, $+jX = (296\Omega)(\tan 116.9 \text{ degrees} - 90 \text{ degrees}) = +j150\Omega$.

The electrical length of the 11.25ft counterpoise wire at 21.1MHz, has a wavelength of $(984)/(21.1MHz) = 46.64$ft, and is $(11.25ft/46.64ft)(360 \text{ degrees}) = 86.84$ degrees. With this electrical length the capacitive reactance of one counterpoise wire at 21.1MHz is, $-jX = (Zo)/(\tan\Theta) = (397\Omega)/(\tan 86.84 \text{ degrees}) = -j21.9\Omega$. Since the two counterpoise wires are in parallel at the feed point, the total capacitive reactance of the two counterpoise wires is, $(-j21.9)/(2) = -j11\Omega$ at 21.1MHz .

Adding the inductive reactance of the monopole to the capacitive reactance of the counterpoise wires will give a total monopole feed point reactance of $(+j150\Omega)+(-j11\Omega) = +j139\Omega$.

Using the formula to estimate the radiation resistance of a 116.9 degree monopole, Rr = 97Ω. For a total calculated feed point impedance of, $97+j139\Omega$. Software calculated, $Z = 89.27+j190\Omega$.

Using the calculated inductive reactance of $+j139\Omega$, the required capacitance to cancel the feed point inductive reactance will be, $C = 1/(2\pi)(139\Omega)(21.1\text{x}10^6) = 54.3$pF. With 54.3pF added in series at the feed point of the monopole, on 21.1MHz, software calculated $Z = 89.21-j11.27\Omega$, 1.83:1 SWR.

A 3-band monopole, NO perfect ground required, and all three frequencies with an SWR less than 2.5:1 using a counterpoise of two 11.25ft wires 1ft above ground!

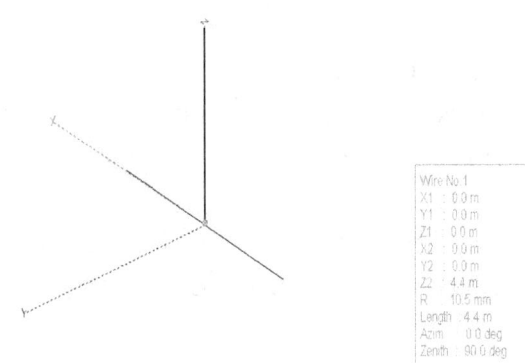

Wire No.1
X1 : 0.0 m
Y1 : 0.0 m
Z1 : 0.0 m
X2 : 0.0 m
Y2 : 0.0 m
Z2 : 4.4 m
R : 10.5 mm
Length : 4.4 m
Azim : 0.0 deg
Zenith : 90.0 deg

4.3 Reducing Counterpoise Length Using Inductive Loading

In a previous section it was shown that a larger diameter conductor reduced the required length of the counterpoise. Another way to reduce the counterpoise conductor length is to use inductive loading.

In this example I will reduce the length of the counterpoise wire using loading coils. As with inductive loading of monopoles the use of coils in the counterpoise will result in increased loss due to the resistance of the wire in the coil. A trade-off for reduced length.

Previously in section 4.1 it was calculated that two 11.3ft (3.44m) wires as single wire transmission lines, forming a counterpoise, would produce the capacitive reactance necessary to cancel the inductive reactance at the feed point of the 50Ω, 18.1MHz monopole.

Let's reduce the two counterpoise wires from 11.3ft, to 6ft (1.83m) each.

In section 4.1 it wS found that 11.3ft was 75.05 degrees we need to find the length, in degrees of the 6ft wire. The length in degrees of the 6ft wire will be (6ft/11.3ft)(75.05 degrees) = 39.85 degrees by proportions. Another way is (6ft/54.37ft)(360 degrees) = 39.73 degrees, where 54.37ft is the length of a wave at 18.1Mhz.

The capacitive reactance of 39.85 degrees of Zo = 397Ω characteristic impedance single wire transmission line will be, -jX = (397Ω)/(tan 39.85 degrees) = -j476Ω. A single 6ft wire provides -j476Ω. We need -j106Ω from each counterpoise wire. The -j476Ω of the 6ft wire has more capacitive ractance than needed. We cancel the excess capacitive reactance, 476Ω-106Ω = -j370Ω by adding an inductive reactance of +j370Ω to each 6ft counterpoise wire at the feed point.

The inductance of +j370Ω reactance at 18.1MHz will be, L = (370Ω)/(2π18.1) = 3.25µH. With a 3.25µH inductor added to each of the two 6ft counterpoise wires, at the feed point, aT 18.1MHz, software calculated, Z = 46.81+j6.429, 1.16:1 SWR.

This is remarkable. With the full length, 11.3ft counterpoise wires software calculated Z = 49.79-j3.445, 1.07 SWR. Though the counterpoise wire length has been reduced by nearly 50%, the radiation resistance reduced by 6%.

Keep in mind the trade-off. The coils at the high current feed point will add resistive loss not included in the software calculation, which is set for no loss. High Q inductors would lower coil losses.

To reduce resistive loss in the counterpoise inductors, the counterpoise wires can be center loaded as was done in the section "Center vs Feed Point Inductive Loading" of the monopole. In that section we developed a short cut for inductive center loading of a monopole, double the required base loading inductance for center loading.

Doubling the 3.25µH base loading inductors to 6.5µH, and placing this inductance in the centers of the two 6ft counterpoise wires, software calculated Z = 49.32-j7.233Ω, 1.16:1 SWR. Our center loading short cut works for counterpoise wires too.

Now, let's change the antenna and counterpoise height above ground in software, its very easy to do.

Software calculated, with the two 3.25µH at the feed point, Z = 46.81+j6.429Ω for a height of 12 inches above ground, the original design height.

Doubling the height to 2ft above ground software calculated, Z = 42.93 -j14.64, 1.42:1 SWR.

Doubling again to 4ft above ground, Z = 37.56-j31.1, 2.13:1 SWR.

Doubling once more, to 8ft, Z = 29.95-j41.26, 3.08:1 SWR.

Even though we have removed the monopole from ground with the counterpoise, we didn't remove the ground image that creates the capacitance the counterpoise relies upon. Notice that for each increase in height the capacitive reactance becomes larger. Raising the counterpoise increases spacing and reduces its capacitance, and due to the inverse relationship between capacitance and capacitive reactance -jX increases.

We can fix the -jX change by recalculating counterpoise Zo and wire lengths for the single wire counterpoise transmission lines at any height above ground we select. This also shows that ground plane verticals using above ground (so called elevated) radials still depend on height above ground.

The radiation resistance is reducing toward a free space value of around 23Ω.

4.4 Capacitive End Loading the 50 Ohm Counterpoise Monopole

In this example the 14.4ft (4.30m) conduit vertical will be reduced to 10ft (3.048m), while still using the inductively loaded two-wire counterpoise. Capacitive end loading at the top will be used to bring the shortened antenna back to resonance at 18.1MHz.

Two wire T end loading will be used at the top to bring the antenna back to resonance, forming an "I" antenna with the two wire counterpoise.

Reducing the physical length from 14.4ft to 10ft reduces the electrical length to (10ft/14.4ft)(100.5 degrees) = 69.79 degrees. In electrical length the antenna is now missing (100.5 degrees − 69.79 degrees) = 30.71 degrees. Or, (4.4ft/14.4ft)(100.5 degrees) = 30.71 degrees for the missing length. Either way you calculate it, we need to replace this missing 30.71 degree electrical length at the top of the conduit monopole with capacitive end loading.

The antenna characteristic impedance for a 14.4ft conduit monopole is,
$Za = 60[Ln(2)(14.4ft)(12in/ft)/(0.922in) -1] = 296Ω$. The capacitive reactance of the missing 4.4ft can be found using basic transmission line theory, $-jX = (Za)/(\tan\Theta) = (296Ω/\tan 30.71 degrees) = -j498Ω$.

For two wire T top loading, each of the two wires will need to have a capacitive reactance twice -j498Ω, -j996Ω. Equal length will insure the current will be half in each wire to create equal but opposite currents in the two wires. This balance will cancel radiation from the horizontal top loading wires, just as it did for the bottom loading counterpoise wires.

The characteristic impedance of a single wire transmission line at 11ft above ground (3.35m), 10ft for the conduit monopole height, plus 1ft above ground, using wire is, $Zo = 60Ln(4)(11ft)(12in/ft)/(0.0641in) = 541Ω$. Now to find the electrical length in degrees of a single wire transmission line with a Zo of 541Ω, to provide -j984Ω of capacitive reactance. The electrical length, $\Theta = \tan^{-1}(Zo)/(-jX) = \tan^{-1}(541Ω)/(996Ω) = 28.51$ degrees.

To find the physical length we need the length of the wave at 18.1MHz, which will be, (984)/(18.1MHz) = 54.37ft.

The physical length of the wire will be, (28.51degrees/360 degrees)(54.37ft) = 4.31ft. Each of the two wires forming the flat top end loading capacitance needs to be 4.31ft (1.31m). With the two 4.31ft horizontal wires added to the top end of the 10ft conduit monopole, with the inductively loaded shortened counterpoise wires at the base, from the previous example, software calculated, $Z = 35.06-j3.872\Omega$, 1.44:1 SWR. The estimated $Rr = 36.6\Omega(\sin^2 69.79$ degrees$) = 32.2\Omega$.

The additional shortening reduced the feed point resistance from near 50Ω to 35Ω. This still gives a very useful SWR for a short antenna.

The feed point resistance can be brought back closer to 50Ω by increasing the electrical length beyond 100.5 degrees. This will be done in the next example.

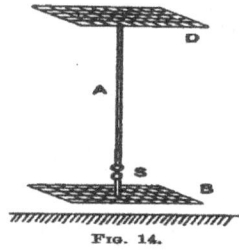

FIG. 14.

Here is a 1916 version of an antenna from J. A. Fleming's "An Elementary Manual of Radiotelegraphy and Radiotelephony for Students and Operators" of a non-earthed monopole antenna using a counterpoise and capacitive top loading. Location "S" is where the spark transmitter would be connected.

4.5 An End Loaded, Bottom Counterpoise, 50 Ohm Monopole

This monopole example will use many of the resources that have been developed up to this point. I will design a top end loaded, 50Ω radiation resistance, monopole for 14.1MHz with a two-wire counterpoise.

In the previous example of the end loaded monopole with inductively loaded counterpoise, the radiation resistance was less than 50Ω. The final radiation resistance, according to software, was around 35Ω. That monopole still had an acceptable SWR.

To show how easy it is to raise the radiation resistance of the monopole antenna I will include an electrical length adjustment in this example. I chose 110 total electrical degrees (instead of 100.5 degrees) which should give a radiation resistance of about 75Ω as a starting point (I found this electrical length using the Rr estimation formula), aiming for around 50Ω as the final radiation resistance after top and bottom capacitive loading have been added.

I will also examine several variations of capacitive loading, both for the counterpoise and the top.

First, calculate the physical length of the 110 degree monopole, (110 degrees)/(90 degrees) = 1.22, resulting in (234)(1.22) = 286, and (286)/(14.1MHz) = 20.3ft (6.19m).

Now to calculate the feed point inductive reactance treating the antenna as transmission line. The characteristic impedance of the antenna as a transmission line will be, $Za = 60[Ln(2)(20.3ft)(12in/ft)/(0.922in) -1] = 316\Omega$.

The antenna is 110 degrees long. The upper 90 degrees is an inverter providing a short to the bottom 20 degrees (110degrees - 90 degrees) of transmission line. The inductive reactance of a shorted transmission line of 20 degrees, calculated from, $+jX = (Za)(tan\Theta) = (316\Omega)(tan\ 20\ degrees) = +j115\Omega$. For a 20.3ft conduit monople over perfect ground, software calculated $Z = 76.66+j118.2\Omega$.

In my real world I don't have a perfectly conducting ground plane and can't cancel the feed point inductive reactance with a physical capacitor. Instead I will use a capacitve counterpoise to provide $-j115\Omega$ of capacitive reactance. I will use a two wire counterpoise for balance. I want equal currents in the two wires for current cancellation and no radiation. To accomplish this the reactance of each wire must be twice $-j115\Omega$, or $-j230\Omega$ for the current to be half the total in each counterpoise wire.

The counterpoise and base of the monopole will be1ft (0.305m) above ground. The characteristic impedance of a horizontal wire 1ft above ground will be $Zo = 60Ln(4)(12in)/(.0641in) = 397\Omega$.

To find the electrical length of the single wire open transmission line with 397Ω characteristic impedance required for $-j230\Omega$ of reactance, $\Theta = tan^{-1}(Zo/-jX) = tan^{-1}(397\Omega)/(230\Omega) = 59.91$ degrees. To find the physical length of the required wire from the electrical length we need to find the length of a wave at 14.1MHz. The length of the wave is, (984/14.1MHz) = 69.8ft. The physical length will be, (59.91 degrees/360 degrees)(69.8ft) = 11.6ft, for each of the two counterpoise wires.

With the two 11.6ft (3.54m) wires added at the base of the 110 degree, 20.3ft, conduit monopole raised 1ft above ground, software calculated, $Z = 65.89-j12.45\Omega$, 1.42:1 SWR.

I don't want a 20.3ft monopole. I want something shorter and capacitively top loaded. Looking in the conduit pile I find I can do a 15ft (4.57m) conduit radiator with what is on hand. I will replace the missing 5.3 feet (20.3ft – 15ft) with capacitive end loading at the top. The missing 5.3ft (1.62m) represents (5.3ft/20.3ft)(110 degrees) = 28.72 degrees.

Now to find the capacitive reactance of the 5.3ft, 28.72 degree, missing conduit length. This will be found from the basic transmission line formula for an open transmission line, $-jX = (Za)/(tan\Theta)$. We know $\Theta = 28.72$ degrees and the conduit mast has $Za = 316\Omega$, calculated previously. The missing capacitive reactance of 28.72 degrees of the conduit monopole will be, $-jX = (Za)/(tan\Theta) = (316\Omega)/(tan\ 28.72\ degrees) = -j577\Omega$.

A two wire capacitive end load to create a T antenna (more of an "I" antenna with the two wire counterpoise) will be used. Each wire will need to have a capacitive reactance twice $-j577\Omega$, or $-j1154\Omega$ per wire. The top wires will be 16ft (4.88m) above ground, the 15ft height of the monopole and the 1ft the monople is above ground. Calculating, $Zo = 60Ln(4)(16)(12in/ft)/0.0641 = 564\Omega$ for the top loading wires.

As was done for the bottom counterpoise wires, each wire will need a reactance of (2)(564\Omega$), $\Theta = tan^{-1}(564\Omega/1154\Omega) = 26.05$ degrees of single-wire transmission line for each top loading wire. The physical length will be, (26.05 degrees/360 degrees)(69.8ft) = 5.05ft (1.54m) for each of the two upper end loading wires.

Software calculated, $Z = 54.74-j16.38\Omega$, 1.38:1 SWR, for two 5.05ft capacitive end loading wires added to the top of the 15ft conduit vertical with the two 11.6ft counterpoise wires at the feed point.

Let's see what length conduit, rather than wire, could be used to top load the antenna. The conduit top loading would be self supporting, unlike wire.

First find the Zo of the conduit 16ft above ground, $Zo = 60Ln(4)(16ft)(12ft/in)/(0.922in) = 404\Omega$. Since the conduit has more surface area than the wire it will have more capacitance. The required length of conduit for capacitive end loading will be less than for the smaller diameter wire.

As before, each length of conduit must have a capacitive reactance of -j1154Ω. For a Zo = 404Ω we need two lengths of conduit that will provide -j1154Ω each. The required electrical length of conduit will be, Θ = tan⁻¹ (404Ω)/(1154Ω) = 19.29 degrees. The physical length will be (19.29 degrees/360 degrees)(69.8ft) = 3.74ft. That will be 3.74ft for each conduit.

Software calculated, Z = 59.28+j0.7471Ω, 1.19:1 SWR, using two 3.74ft conduits for top end loading in place of the wire, with the two 11.6ft (3.54m) counterpoise wires at the bottom.

Since the top loading conduit, unlike wire, requires no end support, perhaps I could hang bird houses or bird feeders from them as cover for a 20M stealth antenna?

I have explored two possibilities for top loading using wires and conduits.

Now to the counterpoise. Thus far lengths of wire have been used. The two 11.6ft wires 1ft above ground may not fit. One way to reduce the counterpoise length would be to use inductive loading. Before trying inductive loading of the wire let's replace the counterpoise wires with conduits.

The characteristic impedance of conduit 1ft above ground will be, Zo = 60Ln(4)(12in)/ (0.922in) = 237Ω. As with top loading, the larger diameter conduit will have greater surface area and therefore more capacitance.

Previously we calculated that the monopole had +j115Ω inductive reactance at the base to be canceled by -j115Ω of capacitive reactance provided by the two wire counterpoise. As with the wire example each of the two conduit conductors must provide -j230Ω for a balanced counterpoise. The electrical length will be, Θ = tan⁻¹ (237Ω)/(230Ω) = 45.86 degrees. The physical length will be (45.86 degrees/360 degrees)(68.9ft) = 8.89ft (2.71m).

With two 8.89ft lengths of conduit for the counterpoise, and conduit top end loading, software calculated, Z = 58.04+j2.764Ω, 1.17:1 SWR.

Let's do something that has not been done before, use four equal lengths of conduit, all at right angles for the counterpoise to provide the required -j115Ω.

In the case of four counterpoise conductors, to keep current balance, each will need to provide four times the required total reactance, (4)(-j115Ω) = -j460Ω each, so the current will divide equally into fourths.

First find the length in degrees, $\Theta = \tan^{-1}(237\Omega)/(460\Omega) = 27.26$ degrees. The physical length will be, (27.26 degrees/360 degrees)(69.8ft) = 5.29ft (1.61m) each.

Using four conduits as the counterpoise, 5.29ft each, at right angles, and with the two conduit top loading, software calculated, $Z = 55.73-j4.633\Omega$, 1.15:1 SWR.

If wires 1ft above ground will work for your situation, but a length of 11.6ft each won't, the wires can be shortened using series inductive loading.

First you must decide the length of wires you can accommodate. For this example I will reduce the 11.6ft length by more than 50% and pick 5ft (1.52m).

We need to calculate the capacitive reactance of a 5ft long wire 1ft above ground as a one-wire transmission line using basic transmission line methods. The electrical length of 5ft will be (5ft/69.8ft)(360 degrees) = 25.79 degrees.

Using this electrical length and the Zo of the wire 1ft above ground, $-jX = Zo/\tan\Theta = (397\Omega)/(\tan 25.79$ degrees$) = -j822\Omega$. Each 5ft wire has too much capacitive reactance by $(822\Omega-230\Omega) = -j592\Omega$. The -j592 can be canceled with an inductive reactance of $+j592\Omega$ in series with each of the two 5ft counterpoise wires at the feed point. The required inductance will be, $592\Omega/2\pi(14.1) = 6.68\mu H$.

With $6.68\mu H$ inductors connected in series with each of the two 5ft counterpoise wires, at the feed point, software calculated, $Z = 54.37+j12.13\Omega$, 1.28:1 SWR. As we know, the 5ft counterpoise wires can be center loaded using twice the inductance needed at the feed point which will be $13.4\mu H$.

There you have it, several possibilities to select from for bottom counterpoise and top end loading, to produce a 14.1MHz, less than 1/4-wave, 50Ω monopole with 15ft of vertical height conduit, 16ft total height above ground level. All done with a calculator, using the antenna as transmission line model, and confirmed by software.

The cage "T" antenna at amateur radio station 1BCG was constructed by members of the Radio Club of America. The cage T antenna was positioned over a wire counterpoise "screen" above the "shack" roof.

The transmitter at 1BCG operated on 230M, about 1.3MHz. This antenna and counterpoise arrangement was used to send the first CW transatlantic message by amateur radio, December 1921.

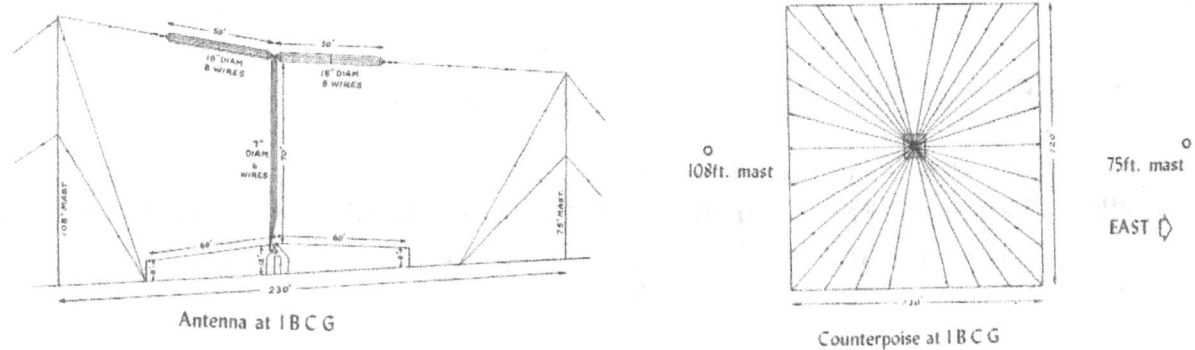

Antenna at I B C G   Counterpoise at I B C G

4.6 The Bottoms-Up Counterpoise 50 Ohm Monopole

Since the monopole has been freed of its earthly tether, thanks to the counterpoise, let's flip it 180 degrees and place the feed point and counterpoise up, creating an inverted monopole.

Why? Perhaps we need a second floor fed antenna. Another reason would be to get the high current feed point away from ground to reduce loss.

I'm going to make the monopole 100.5 degrees long on 21.1MHz for a nominal 50Ω feed point radiation resistance. Since this antenna is longer than 90 degrees the reactance at the feed point will be an inductive (+jX) reactance.

The length formula for a 100.5 degree monopole is, (100.5degrees/90degrees)(234) = 261/f. At 21.1MHz the length will be, 261/21.1 = 12.4ft. For the 12.4ft (3.78m) I will use conduit for the monopole. The characteristic impedance of the 12.4ft conduit monopole as a transmission line will be, Za = 60[Ln(2)(L)/(d) -1] = 60[Ln(2)(12.4ft)(12in/ft)/(.922in) -1] = 287Ω.

The inductive reactance of the 100.5 degree monopole found using basic transmission line theory, +j = (Za)(tanΘ) = (287Ω)(tan 100.5 degrees – 90 degrees) = +j53.2Ω. The angle Θ is written as (100.5 degrees – 90 degrees) because the 90 degrees from the open end is a 1/4-wave impedance inverter places a short on the remaining 10.5 degrees of the antenna as transmission line.

The 100.5 degree conduit monople will need -j53.2Ω of capacitive reactance at the feed point for resonance on 21.1MHz. That amount of capacitive reactance will be provided by two, equal length, single-wire horizontal transmission lines, for balance and radiation cancellation.

The conduit monopole is 12.4 ft long, and I will add 1 ft to get the opposite end above ground level. This will place the feed point and counterpoise wires of the inverted monopole antenna 13.4ft (4.08m) above ground.

A single wire transmission line has a characteristic impedance of, $Zo = 60Ln(4)(h)/(d)$, where "h" is height above ground and "d" is the conductor diameter, in the same units. For our single wire transmission line counterpoise, 13.4ft above ground $Zo = 60Ln(4)(13.4ft)(12in/ft)/(0.0641in) = 553\Omega$.

To keep the counterpoise balanced and prevent radiation, two parallel connected wires will be used. We need a total of $-j53.2\Omega$, each counterpoise wire will contribute twice the total required reactance, $(2)(-j53.2\Omega) = -j106\Omega$. The electrical length, in degrees, of a single wire transmission line with a $Zo = 553\Omega$, to provide a capacitive reactance of $-j106\Omega$, will be $\Theta = \tan^{-1}(Zo)/(-jX) = \tan^{-1}(553\Omega)/(106\Omega) = 79.15$ degrees.

Now the physical length can be calculated. To do that I need the length of a wave at 21.1MHz, which can be found from $(984/21.1MHz) = 46.64ft$. Converting degrees to feet, $(79.15 degrees/360 degrees)(46.64ft) = 10.25ft$. Two co-linear wires, 10.25ft (3.21m) each should do it. (This could be another opportunity to use conduit because it would be self supporting, unlike the wire. And shorter too.)

Software calculated, for two 10.25ft (3.12m) counterpoise wires, 13.4ft above ground at the feed point, $Z = 48.53+j3.643\Omega$, 1.08:1 SWR at 21.1MHz.

Since we have easy access to the open end of the monopole near ground level, we can add capacitive end loading near the ground for resonance on 18.1MHz and 14.1MHz and produce a three band inverted counterpoise monople antenna.

What follows are the methods using transmission line theory for finding 2-wire capacitive end loading lengths for operation on 18.1MHz and 14.1MHz.

For 18.1MHz. Since the length of each counterpoise wire is fixed at 10.25ft for 21.1MHz, we need to find the capacitive reactance of the counterpoise at 18.1MHz. We already know that the characteristic impedance of the counterpoise wire is 553Ω. We need to find the -jX of the counterpoise for the new wavelength at the frequency of 18.1MHz, which is $(984/18.1MHz) = 54.37ft$.

The length of the counterpoise in degrees at 18.1MHz is now $(10.25ft/54.37ft)(360$ degrees$) = 67.87$ degrees. To find the capacitive reactance of the counterpoise wire at 18.1MHz use the following basic transmission line relationship, $-jX = (Zo)/(\tan\Theta)$ which for our example is, $-jX = (553\Omega)/(\tan 67.87 degrees) = -j225\Omega$ for each wire.

Since the counterpoise reactances are in parallel the total capacitive reactance for the two wires at the feed point on 18.1MHz is, (-j225Ω)(/2) = -j113Ω.

Now that we know the counterpoise capacitive reactance at 18.1MHz presented to the feed point of the conduit monopole is -j113Ω we need to find a new monopole length that will provide +j113Ω at the feed point to cancel the new capacitive reactance of -j113Ω at 18.1MHz. The new length in degrees of a conduit monopole with an inductive reactance of +j113Ω can be found using $\Theta = \tan^{-1}(+jX)/(Za) = \tan^{-1}(113Ω)/(287Ω) = 21.49$ degrees, which when added to the existing 90 degrees will be 111.5 degrees for our new 18.1MHz monopole electrical length.

Now for the new physical length for a 111.5 degree monopole, (111.5 degree/90 degrees)(234) = 290, and, (290/18.1MHz) = 16ft (4.88m) for the physical length of conduit needed at 18.1MHz.

We may need 16ft, but what we have is the 21.1MHz length conduit of 12.4ft. We are missing 3.6ft (16ft-12.4ft) which will be replaced by 2-wire capacitive end loading wires 1ft above ground.

The missing length of conduit in degrees, (3.6ft/16ft)(111.5 degrees) = 25.09 degrees. The missing 25.09 degrees of conduit will have a capacitive reactance of, $-jX = (287Ω)/(\tan 25.09$ degrees) = -j613Ω total capacitive reactance to be replaced by end loading.

The end loading will be provided by two wires 1ft above ground Zo = 60Ln(4)(12in)/0.0641in = 397Ω. The total capacitive reactance required is -j613Ω, but we will be using 2-wire capacitive loading to have balance for no horizontal radiation, so each wire must have a reactance of 2(-j613Ω) = -j1226Ω. The electrical length of wire required will be, $\Theta = \tan^{-1}(397Ω)/(1226Ω) = 17.94$ degrees, for a physical length of (17.94degrees/360degrees)(54.37ft) = 2.71ft.

Using two 2.71ft wires (0.826m), 1ft above ground for capacitive end loading added to the 21.1MHz inverted counterpoise (now an "I") conduit monopole, software calculated, Z = 57.09+j0.1851Ω, 1.14:1 SWR on 18.1MHz.

If you have followed all the twists and turns to this point, I congratulate you and applaud your calculator operational skills!

If you took the basic 21.1MHz bottoms-up monopole software model and used it to "successive approximate" or "optimize" this antenna for 18.1MHz to find the wire lengths for end loading, I certainly do understand why you did.

Even though the solution for operation on 14.1MHz might more easily and sooner done by successive approximating with the software, here goes.

For 14.1MHz. Find the reactance of the 10.25ft counterpoise wires for 14.1MHz. To do that we need the physical length of the wave, $(984)/(14.1) = 69.8$ft. The electrical length will be, $\Theta = (10.25)/(69.8)(360$ degrees$) = 52.87$ degrees. The capacitive reactance of the 52.87 degrees of the 553Ω impedance of counterpoise wire will be $-jX = (Zo)/(\tan\Theta) = (553\Omega)/(\tan 52.87$ degrees$) = -j419\Omega$ each, for a total of $(-j419\Omega)/(2) = -j210\Omega$ for the counterpoise wires.

To cancel the $-j210\Omega$ capacitive reactance of the counterpoise at the feed point we need a monopole length with an inductive reactance of $+j210\Omega$. To get a $+j210\Omega$ with a 287Ω characteristic impedance transmission line of conduit, $\Theta = \tan^{-1} (+jX/Zo) = \tan^{-1} (210\Omega)/(287\Omega) = 36.19$ degrees. For the total new monopole length we add 90 degrees, $(36.19 +90$ degrees$) = 126.19$ degrees, which I will call 126.2 degrees.

The physical length of the monopole required will be, $(126.5$ degrees$/90$ degrees$)(234) = 329$ and $(329/14.1$MHz$) = 23.3$ft (7.10m). Since the conduit is 12.4ft long it is missing 10.9ft (3.32m) of length (23.3ft-12.4ft) at 14.1MHz. The missing 10.9ft represents, $(10.9$ft$/23.3$ft$)(126.2$ degrees$) = 59.04$ degrees. The missing 59.04 degrees of conduit will have a capacitive reactance of, $-jX = (287\Omega)/(\tan 59.04$ degrees$) = -j172\Omega$, which will be replaced by 2-wire end loading.

The total capacitive end loading required is $-j172\Omega$, to create a balanced end load two wires will be used, each will need to provide twice the reactance of $-j172\Omega = -j344\Omega$ for each wire. The electrical length of the wire, 1ft above ground to provide $-j344\Omega$ end loading can be found as follows, $\Theta = \tan^{-1} (397\Omega)/(344\Omega) = 49.09$ degrees. The physical length will be $(49.09$ degrees$/360$ degrees$)(68.9$ft$) = 9.52$ft (2.90m).

Using two 9.52ft (2.82m) co-linear wires at the bottom end of the 21.1MHz inverted conduit monopole, 1ft above ground, at 14.1MHz, software calculated $Z = 59.64+j24.54\Omega$, 1.61:1 SWR.

A 12.4ft conduit with the feed point at the top, and using two 10.25ft, co-linear wires to form a counterpoise at the feed point creates an inverted vertical monopole for 21.1MHz.

By adding wire at the opposite end, near ground, it becomes possible to use the same inverted antenna on 17M and 20M simply by adding capacitive end loading wires, 1ft above ground level. Another option for 14MHz operation is to add an inductance between the bottom end of the conduit and the two 2.71ft wires (0.826m) for 18.1MHz.

Getting the feed point above ground level moves the high current point of this monopole further away from ground loss.

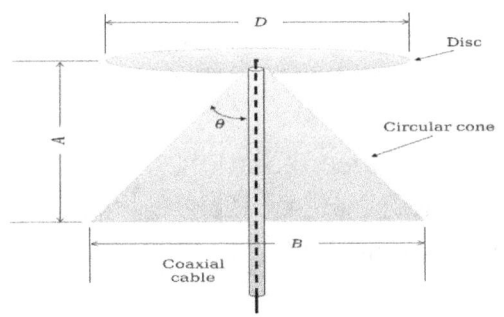

The "Discone" antenna (left) inverts the Marconi conical antenna and uses a disk as the counterpoise at the top of the antenna. The feed point is elevated above ground.

4.7 An Inductive Counterpoise

Thus far, the reason I concentrated on making the monopole longer than 90 degrees, which caused an inductive reactance at the feed point, was to raise the feed point radiation resistance of the monopole above 36.6Ω and provide a better match for a 50Ω source. The inductive reactance at the feed point of the longer than 1/4-wave monopole also allowed for its removal from the impossible perfect ground by application of a capacitive ground, the counterpoise.

Just as a monopole longer than 1/4-wave (90 degrees) will have an inductive reactance at its feed point, so will a longer than 90 degree counterpoise wire. By making the wires for the counterpoise longer than 90 degrees a shorter than 90 degree monopole, with capacitive reactance at its feed point, can be brought to resonance with an inductive counterpoise system.

Here is an example of how that can be accomplished using two counterpoise wires 1ft (0.31m) above ground with a less than 1/4-wave monopole.

I'm going to transmission line model a 15ft (4.572m) conduit monopole for operation at 14.1MHz. The self resonant frequency of the 15ft conduit will be (234/15ft) = 15.6MHz. Being resonant at a frequency higher than 14.1MHz this monopole is short and will have a capacitive reactance at the feed point. The characteristic impedance of the 15ft conduit monopole will be, $Za = 60[Ln(2L/d) - 1] = 60[Ln(2)(15)(12in/ft)/(0.922in) - 1] = 298\Omega$.

The electrical length of the monopole at 14.1MHz will be, (14.1MHz/15.6MHz)(90 degrees) = 81.35 degrees. The reactance at the monopole feed point will be -jX = (298Ω)/(tan 81.35 degrees) = -j45.3Ω.

The total *inductive* reactance a 2-wire counterpoise must provide at the feed point is +j45.3Ω.

Because two counterpoise wires will be used for balance to reduce radiation, the inductive reactance of each counterpoise wire will need to be (2)(+j45.4Ω) = +j90.8Ω. The two wires of the counterpoise make a parallel connection at the base feed point of the short monopole.

The characteristic impedance of a wire 1ft (0.305m) above ground is Zo = 60[Ln(4)(h)/(d)] where "h" is height above ground and "d" is the conductor diameter. Zo = 60[Ln(4)(12in)/(0.0641in)] = 397Ω.

Since part of the Zo of the counterpoise wire will be a 90 degree (1/4-wave) transformer, we need to find the capacitive reactance required at the open end of the 1/4-wave transformer with a Zo = 397Ω (the counterpoise wire) to provide +j90.8Ω at the opposite end of the counterpoise wire, the monopole feed point.

In the chapter on the 1/4-wave monopole as an impedance inverter, -jX = (Zo)²/(+jX) = (397Ω)²/(+90.8Ω) = -j1736Ω. The electrical length of a 397Ω line to provide a capacitive reactance of -j1736Ω will be, Θ = tan⁻¹(397Ω)/(1736Ω) = 12.88 degrees. To this 12.88 degrees we add 90 degrees for the 1/4-wave inverting transmission line transformer giving a total electrical length of 102.9 degrees for each counterpoise wire.

To find the physical length of each counterpoise wire we need the length of a wave at 14.1 MHz, which is (984)/(14.1MHz) = 69.8ft (21.3m). The physical length of each counterpoise wire will be (102.9 degrees/360 degrees)(69.8ft) = 19.95ft (6.08m)

For two 19.95ft (6.08m) counterpoise wires connected to the feed point of the 15ft conduit monopole,1ft above ground, at 14.1MHz, software calculated Z = 25.33-j18.01Ω, 2.33:1 SWR. According to software, with the wires lengthened to 6.4m each, Z = 25.48+j0.2395Ω, 1.96:1 SWR at the feed point.

The estimated radiation resistance of an 81.35 degree monopole over a perfect ground plane will be Rr = 36.6Ω[(1-cosΘ)²/(sin²Θ) = 27Ω. For the 15ft conduit monopole over perfect ground at 14.1MHz, software calculated, Z = 25.52-j50.43Ω, as compared to Z = 27-j45.3Ω from Rr estimation formula and reactance by the transmission line model.

As this example shows, once again, inductive loading at the feed point of a short monopole does nothing to change the radiation resistance. But, in this example, there is no loading coil loss.

Chapter 5 The Horizontal Dipole

In this chapter the examples look at the horizontal dipole and its various derivatives.

The effect of dipole height above earth's surface is examined in this section. It is important to understand that horizontal dipole height above ground plays a role in radiation resistance and characteristic impedance due to the ground image. Any horizontal wire above ground will be affected by the earth's image, as we saw with counterpoise and top loading conductors.

Examples for loading short dipoles using inductive loading, or capacitive end loading by bending the ends are shown. Variations in feed point location ("Windom" or Off-Center Fed) and harmonic operation are examined. As well as the linear horizontal dipole, the inverted V is also examined.

Two frequency operation using a trap is presented. The trap design is not done in the usual way. A different approach I have developed for a double resonant dipole makes it possible to calculate the parallel circuit inductance and capacitance from the antenna Q, and requires only one trap placed at the dipole feed point.

Two dipoles longer than the self resonant half wave length, the Extended Double Zepp (EDZ) and the G5RV, are presented as examples of antennas with increased gain and directivity compared to a half wave dipole.

5.1 The Dipole, Two Monopoles in Series – The Image Becomes Real

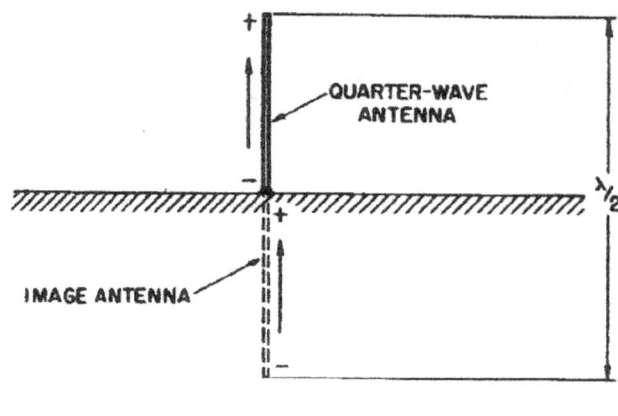

QUARTER-WAVE ANTENNA

IMAGE ANTENNA

$\lambda/2$

TM 666-III

Figure 107. Quarter-wave antenna connected to ground.

Using the tape measure and mirror it was possible to see the monople had another half below the reflecting plane as an image.

Since the reflected image is an exact representation of the monopole (or tape measure), the image can be replaced by another exact representation, another monopole, placing the two monopoles in series at their feed points forms a dipole.

Using the 10ft conduit monopole, over perfect ground, software calculated the impedance at 23.4MHz, the resonant frequency, as, $Z = 36.11+j0.342\Omega$. For a 20ft conduit dipole (two 10ft monopoles in series), center fed in free space, software calculated, at 23.4MHz, the 1/2-wavelength resonant frequency, $Z = 72.17+j0.7072\Omega$.

If the dipole is two monopoles in series, the radiation resistance and reactance of the dipole should be twice that of the monopole, because they are series connected. The radiation resistance of the monopole by software was calculated as 36.11Ω. Twice that radiation resistance will be 72.22Ω. The radiation resistance of the dipole calculated by the same software is, 72.17Ω (the same to 3-digits). Software seems to confirm the series monopole concept.

What happens to any feed point reactance the antenna may have?

For the 10ft conduit as a monopole at 21.1MHz, software calculated, $Z = 25.52-j44.1\Omega$. For the 20ft conduit dipole at 21.1MHz, software calculated, $Z = 51.86-j89.15\Omega$. Twice the radiation resistance of the monople at 21.1MHz, $Rr = 51.04\Omega$.

Now check the reactive component.

Software calculated the reactance of the monopole as, -j44.1Ω. Twice -j44.1Ω of the monopole is -j88.2Ω. Software calculated -j89.15Ω for the dipole at 21.1MHz in free space.

By thinking of a dipole as two series connected monopoles, all the examples in the earlier sections for monopoles over perfect ground can become dipoles in free space by using two of them connected at their feed points. The resistance and any reactance values of the monopole examples will be doubled for the dipole because the two monopoles form a series circuit.

For a dipole at resonance, if you split the dipole anywhere along its length, the two reactances found using transmission line theory will always be equal but opposite, giving a total reactance of zero. (As we will see later, with off-set feed the feed point resistance will be higher, but, the reactance will be zero at resonance everywhere along the dipole.)

As an example, lets split a wire dipole at 30 degrees one side of center. Now we have two transmission line pieces. One will be 60 degrees long (90-30) the other will be 120 degrees long (90+30). For this example I'll use Za = 1000Ω.

The capacitive reactance of the 60 degree part will be, -jX = Za/(tan 60 degrees), -jX = 1000Ω/tan(60 degrees) = -j577Ω, looking out to the open end . Looking toward the other end we see a 120 degree open line. The outermost 90 degrees acts as an inverter, placing a short on the remaining 30 degrees. The inductive reactance of this shorted 30 degree transmission line will be, +jX = Za(tan30 degrees) = 1000Ω(tan30) = +j577.

This little exercise shows that when a dipole antenna is operating at self resonance it provides its own conjugate match with the other part, anywhere you want to slice it.

It may come as no surprise by now that in all formulas used for estimating the Rr of a monopole, where 36.6Ω is indicated, substitute twice 36.6Ω = 73.2Ω for the Rr of a dipole in free space.

In the formula for finding the characteristic impedance of a dipole antenna as a transmission line in free space, substitute 120 for the multiplier 60 used in the monopole formula. In **free space**, the ideal condition for a dipole, the average characteristic impedance becomes, **Za = 120[Ln(2)(L)/(d) -1]**. For the antenna characteristic impedance equation of the dipole the length "L" is half the dipole antenna length.

In both the dipole and the monopole we are dealing with an opened out transmission line as an antenna model. Opening a coax line out half way, the case for a monopole, and a parallel line all the way for a dipole, does not change the length of the original transmission line as an antenna.

In the literature, as in this section, the free space radiation resistance of 73.2Ω will be used for our ideal dipole model. Using software and free space the Rr of a 20ft wire dipole at 23.4MHz was, 72.09Ω. For 20ft of conduit, software calculated, 72.17Ω. Fortunately for antenna design, the thickness of the radiator has a small effect on the radiation resistance.

As for antenna reactance, thickness matters much more, which is why length and diameter are both considered in the Za antenna as transmission line formulas.

As we have seen, when changing the height above ground of a monopole with horizontal counterpoise wires, software showed changes in the radiation resistance and the reactance at the feed point. This was due to the continued existence of a ground image.

As you will see in the next section, when the dipole moves from the ideal of free space toward its earthly image, radiation resistance and reactance will change too, depending upon the height of the dipole above the ground image.

Height above ground very much affects the horizontal dipole. The effect of the ground image will require an adjustment of the free space antenna as a transmission line formula for average characteristic impedance.

Most models of the dipole are a linear, 180 degree radiator. An "inverted V" dipole will be examined after the linear 180 degree horizontal dipole above ground.

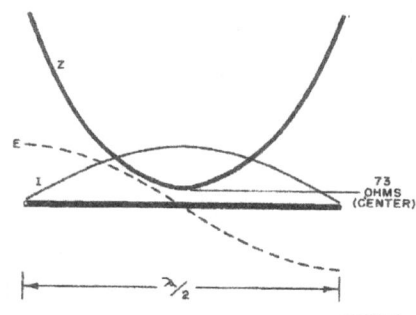

The figure above shows the voltage and current distribution of a resonant 1/2-wave dipole in free space. The "Z", total impedance curve, is minimum at the center of the dipole (marked 73 Ohms), maximum at the open ends.

The "U" shaped impedance (Z) curve of the dipole is similar to that of a low Q, RLC, series circuit with Z minimum equal to radiation resistance, Rr, at resonance.

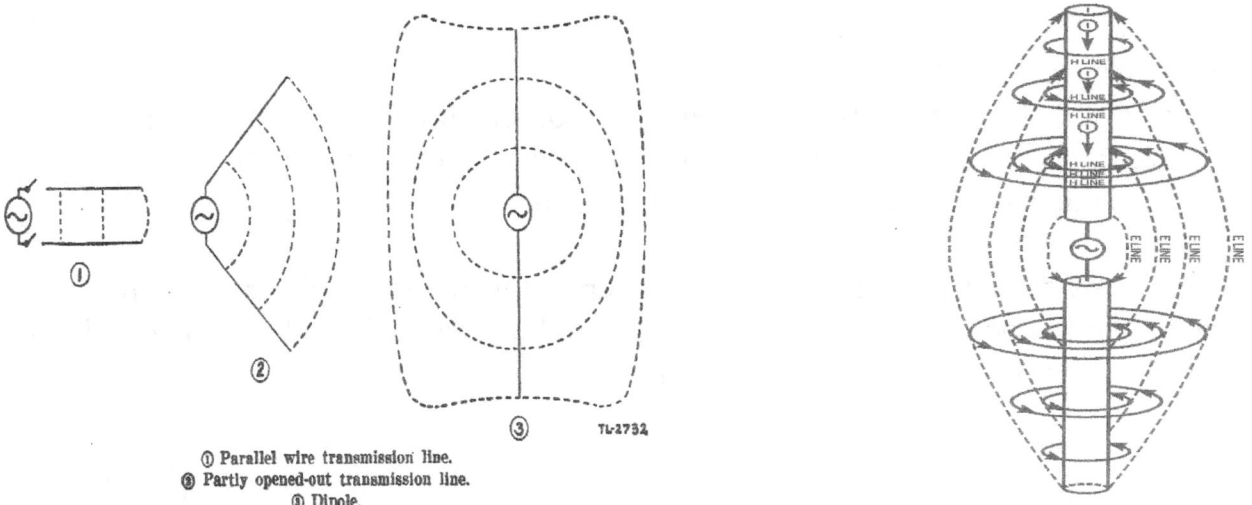

① Parallel wire transmission line.
② Partly opened-out transmission line.
③ Dipole.

The electric (E Line) fields extend from pole to pole of the dipole, while the magnetic fields encircle the dipole conductor.

Being two monopoles in series, the distributed inductance and capacitance will be additive. Inductors in series add to become more inductance. Capacitors in series add inversely, making the total capacitance less.

Since the basic formula for transmission line characteristic impedance is $\sqrt{(L/C)}$, twice as much inductance, L, with half as much capacitance, C, will result in a dipole characteristic impedance twice that of the monopole, causing the multiplier 60 of the monopole to become 120 in the dipole average characteristic impedance formula.

5.2 The Horizontal Dipole Meets Ground

Right away, by calling a dipole horizontal a reference is involved, the earth's surface in the real world. In free space there is no vertical or horizontal for a lone dipole.

If a dipole were not affected by the presence of earth, why not just lay the dipole on the ground and save all the work and aggravation of hanging the thing? Even with no real knowledge about the effects of ground on a horizontal dipole our intuition tells us this is just wrong.

Now that we have established (by reductio ad absurdum) laying the dipole on the surface of the earth may not be a good choice, let's discuss some reasons why we raise our horizontal "ethereal adornments" above ground.

A dipole in space is free of anything that would disturb the natural fields of the dipole. All electric and all magnetic fields are due to the dipole alone. In free space the distributed capacitance and inductance of the antenna, as well as its radiation resistance, will be dependent on the dipole's own self-impedance.

As conductors move from free space toward earth their fields are disturbed by the presence of other objects in its environment, including the earth, and no longer will the dipole characteristics be dependent on itself alone.

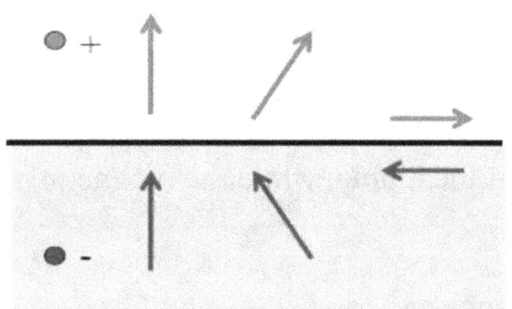
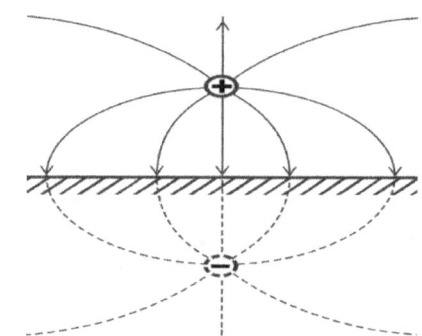

The fields that once were from one half (pole) of the dipole to the other half (pole) begin to have a choice between the antenna itself, the earth below and other objects occupying its space. The dipole, as with any horizontal conductor above ground, will have an earthly image. The dipole is no different than those single wire transmission lines used for capacitive loading to create monopole top loading and counterpoises.

The horizontal conductor direction of current in the image has opposite polarity and direction of current flow. This will be true for the horizontal dipole.

This is exactly how things are supposed to be with any parallel transmission line, the currents flowing in the two conductors are equal but opposite. If this balanced current flow is kept, the transmission line should not radiate, the fields being equal but opposite will cause a net cancellation of the radiation of fields around the transmission line.

This is why the parallel transmission wire line is called a balanced line and should not radiate. And why our two wire counterpoise and the T end loading conductors don't radiate if of equal length, due to field cancellation.

The ground image of a horizontal antenna is trying to balance the current in the dipole and prevent radiation. Fortunately for the dipole antenna (and us) there is not a complete cancellation of fields everywhere.

This is where the height above ground becomes important in the performance of the horizontal dipole as a radiator. Dipole height above ground is a significant factor in radiation resistance, reactance and final radiation pattern.

As the dipole moves from free space toward earth, the field of the dipole and field of the image interact. This interaction, with a time delay related to distance between the dipole and spacing with its image (phase in circuit parlance) will have various possibilities of relationships.

In "B' of the figure below, the reflected ground wave is 180 degrees phase shifted compared to the direct wave.

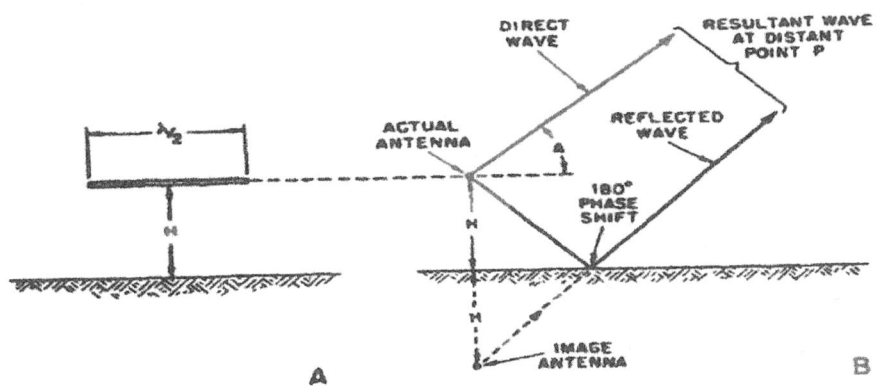

If the delay, due to phase shifts, of the radiated field at some distance is in phase (constructive interference), the fields will be additive. If, on the other hand, the distant fields are out of phase (destructive interference), the fields will subtract. At the extremes of these interference relationships, fields that are completely in phase can sum and become larger. If completely out of phase the fields can result in total cancellation.

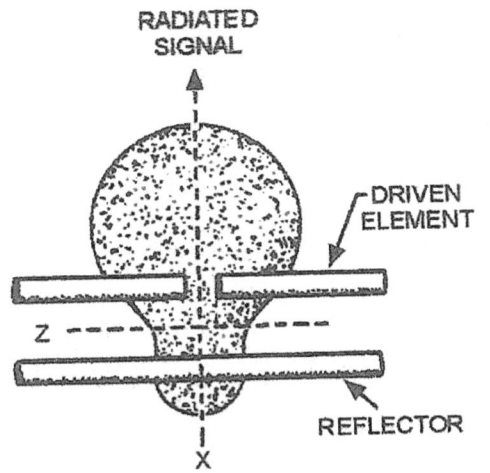

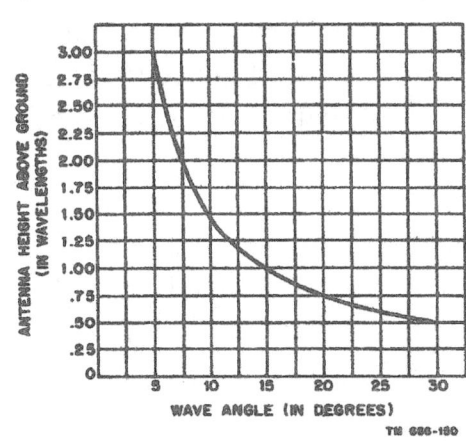

Figure 146. Graph showing variation in wave angle for different antenna heights.

The graph on the right shows the relationship between dipole antenna height above ground and the major lobe "take-off angle." Dipoles must be very high for low angles.

The left figure on the last page shows how a dipole (driven element) above ground (dotted line) and the image (reflector) is more of a two element parasitic beam than a single dipole. Low horizontal dipoles over ground tend to beam up over head toward the sky above. The pattern and radiation resistance will depend on the spacing between the dipole and its parasitic reflector image element, and mutual coupling.

The final radiation field of a low dipole is a superposition of the dipole field and the field associated with the ground reflection. Superposition of fields for linear antennas applies because superposition applies to all linear circuits, including transmission lines.

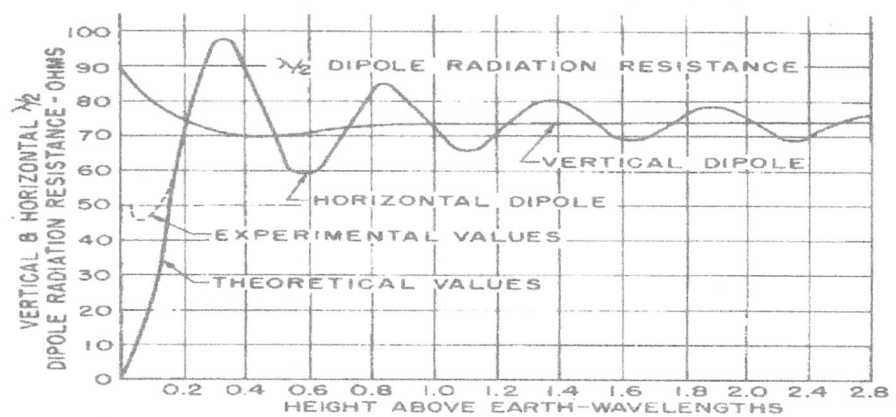

As you move across the graph, above, from left to right, the dipole is (physically and electrically further above ground and the variation in radiation resistance becomes smaller and smaller. If you continued this graph out to infinity it settles at the free space radiation resistance of 73.2Ω, for an infinitely thin conductor. The dipole Rr above ground is a sort of dampened oscillation swinging back and forth around the ideal, free space, radiation resistance of 73Ω as height is varied.

The graph clearly shows the greatest variation in radiation resistance is at low heights, where the opposing image has its greatest influence in causing either constructive or destructive interference in the dipole field. The radiation resistance of the dipole, as a driven element, is influenced by the coupled mutual impedance of its parasitic earthy image.

The graph indicates that the amount of variation with height begins to reduce for heights above about 1/2-wavelength. The dipole at 1/2-wave above perfect ground has a radiation resistance very near the free space ideal. Any dipole below 1/2-wave height will have the greatest variation in radiation resistance. This leads to suggesting a dipole (or horizontal directive array) be at least 1/2-wave above ground.

Notice from the graph that the theoretical curve goes to zero Ohms at zero height. Experimentally measured values are shown by the dashed curved.

The dipole resistance actually reaches a low value around 0.17-wave length (not zero) and then begins to head toward 100Ω resistance at zero height. One source measured the radiation resistance of a dipole lying on the ground to be 120 to 140Ω.

For 0.29 wavelength above ground, about 20ft (6.1m) on 20 meters, it looks like Ro = 96Ω from the graph. Software calculated 91.79Ω for a wire 20M dipole 0.29-wave (6.1m) above ground.

Before computers, the use of such graphs was how things were done, and can still be done. It is still possible to find an estimated radiation resistance for a short dipole antenna above ground using our old friend, $Rr = Ro\ [(1-cos\Theta)^2/(sin^2\Theta)]$ where Ro is found from the graph for the specific height of the short dipole antenna above ground.

Let's work through the steps to find the estimated radiation resistance for a 20ft wire dipole at 14.1MHz, 0.29-wave (6.1m) above ground. The resonant frequency of the 20ft dipole is (468)/(20ft) = 23.4MHz

First the radiation resistance of the 20ft wire dipole, 0.29-wave (6.1m) above ground at 14.1MHz, $Rr = Ro\ [(1-cos\Theta)^2/(sin^2\Theta)]$, where Ro = 96Ω from the graph, and Θ = (14.1MHz/23.4MHz)(90 degrees) = 54.23 degree electrical length, results in an estimated Rr = 25Ω.

Now let's calculate the capacitive reactance of this short antenna at the feed point of the dipole using the antenna as transmission line model. A horizontal dipole antenna over ground can still be treated as a transmission line, the formula for Za will need an adjustment for the presence of the image.

For a **dipole near ground** the average characteristic impedance will be, **Za = 120[Ln(2)(L)/(d) - 0.75]**. The change is the -1 to -0.75 in the formula, the remainder is the same. As always, same units for length and diameter.

Now for the dipole reactance. First we need the antenna characteristic impedance, Za =120[Ln(2)(10ft)(12in/ft)/(0.0641in) – 0.75] = 897Ω. The reactance of the dipole antenna at its center can be found using transmission line theory $-jX = (Za)/(tan\Theta) = (897Ω)/(tan\ 54.23\ degrees) = -j646Ω$.

The final total impedance, as determined with a calculator, using the graph, a formula for estimating Rr, and transmission line theory for a 20ft wire dipole up 0.29-wave (6.1m) at 14.1MHz is, Z = 25-j646Ω. Software calculated, Z = 24.35-j643.9Ω.

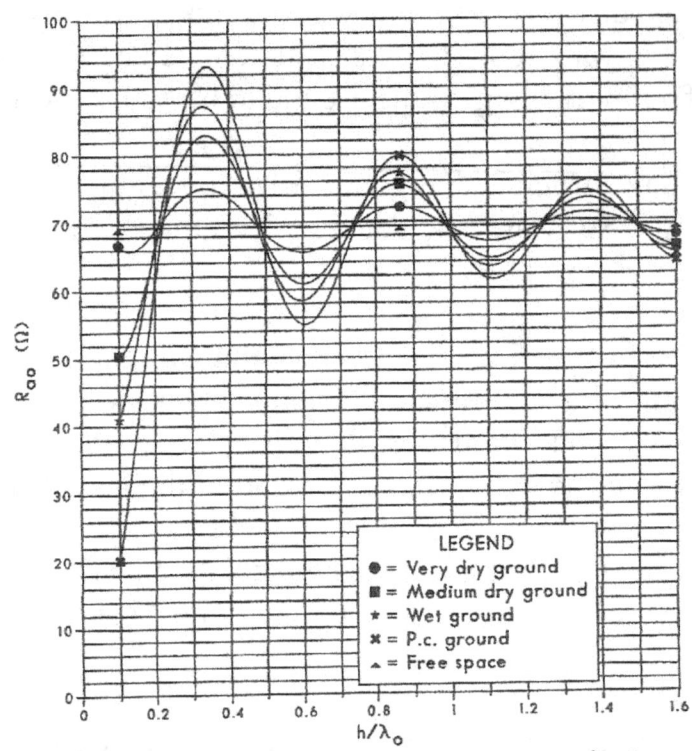

The software and the previous graph agree so closely because both are based on perfect ground. The actual radiation resistance of the dipole will be different dependent on the real ground under the dipole, as shown in this graph from the paper, "Review of Characteristics for HF Dipole Antennas Including the Cases Where the Dipoles are Above and Parallel to the Surface of Real-World Grounds" by G.A.Royer.

For heights less than 0.17-wavelength, the radiation resistance will always be less than 70Ω, even for a full 1/2-wave dipole at resonance. This is worth keeping in mind because many amateur radio dipoles for HF are electrically closer to ground on the lower frequencies, such as 80M and 160M.

The ground image at low heights will also effect horizontal directive antennas. A horizontal gain antenna, such as a Yagi, at low height, can have the gain and elevation angle distorted by the ground image.

The stated gains for horizontal dipole derived directional antennas may not be achieved when close to ground. If you can't get that horizontal Yagi up to 1/2-wave above ground, you likely will not have advertised performance if the stated gains are for a free space model.

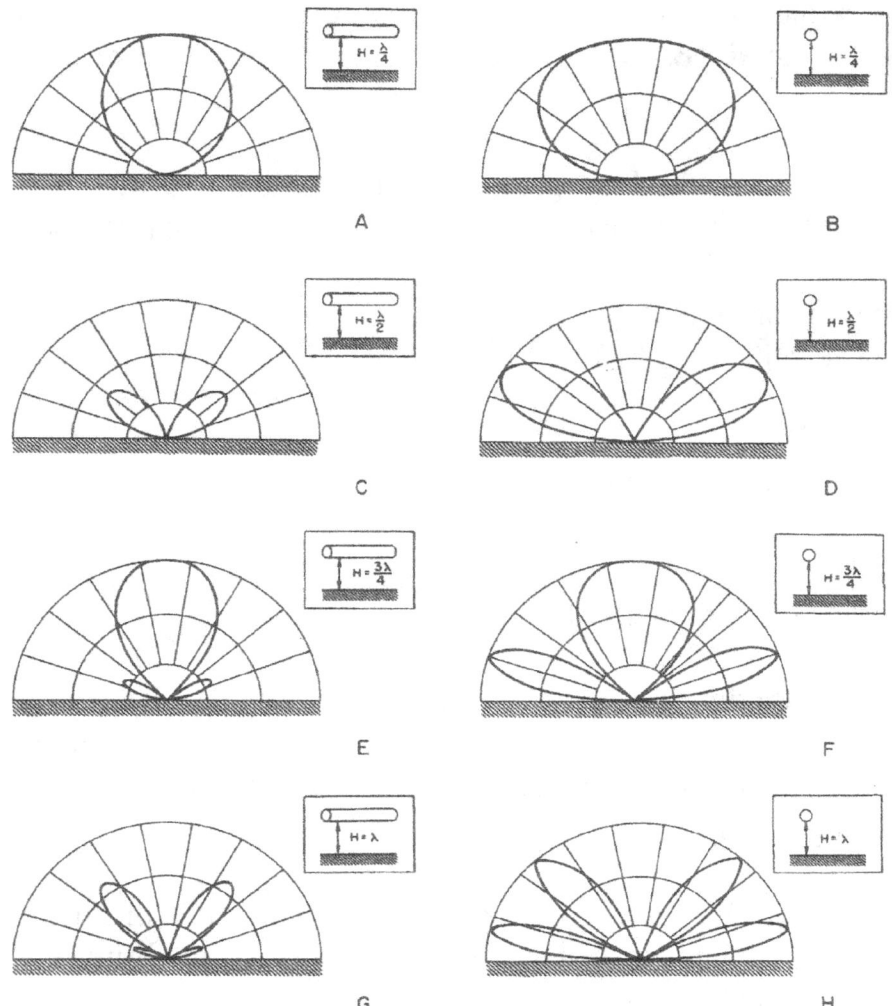

At a height of 1/2-wave and 1-wave length above ground the dipole radiation pattern has no overhead lobe and only low angle lobes.

5.3 The Inverted V Dipole

A popular variation on the linear horizontal dipole antenna is to bend the dipole into a V shape at the feed point, apex up. Let's consider this popular case and the effect of the bend angle on radiation resistance.

The monopole has been viewed as a half opened transmission line, with Rr over a perfect ground as 36.6Ω. The dipole as a fully opened out parallel wire transmission line with a free space Rr of 73.2Ω.

For angles between half and fully open (between 90 and 180 degrees), as will be the case of the inverted V dipole, Rr can be between 36.6Ω and 73.2Ω, depending on the bend angle in free space.

The angle between the dipole poles has an effect on the Rr of an antenna. The relationship is **Rr = Ro[sin(B/2)]²** where B is the bend angle in degrees at the apex, the dipole feed point.

Let's test this new formula for 180 degrees (a straight dipole), 120 degrees (an often recommended inverted V angle) and 90 degrees (the monopole case), using Ro = 73.2Ω.

For a 180 degree (linear dipole),
Rr = 73.2Ω[sin(180/2)]² = 73.2Ω[1] = 73.2Ω.

For a 120 degree "V",
Rr = 73.2Ω[sin(120/2)]² = 73.2Ω[0.75] = 54.9Ω.

For 90 degree "V",
Rr = 73.2Ω[sin(90/2)]² = 73.2Ω[0.5] = 36.6Ω

At 120 degrees the radiation resistance of an inverted V dipole in free space is very close to 50Ω, which is why it is offered as a good choice for an inverted V dipole angle. But, as we shall see, only in free space. Most HF inverted V dipoles are not likely to have heights approaching free space.

I constructed a virtual 20M, 120 degree, wire inverted V antenna in software. For the model in free space, software calculated, Rr = 57.56Ω. Our new formula for a 120 degree bent dipole estimated Rr = 54.9Ω.

Bringing the V dipole down to earth will change the Rr of the inverted V dipole just as it does for a linear dipole, depending upon height above ground. Changing the software model from free space to an apex at 30ft (9.14m) above ground, software calculated, Rr = 73.66Ω.

Usually I do the calculator stuff first, but in this example I am doing it after so you have the software Rr first. There is a point to be made regarding the inverted V dipole compared to a horizontal 180 degree, linear, dipole height above ground and radiation resistance.

If we use the radiation resistance graph at 30ft apex height, which is 0.43-wave above ground for 20M, I see about 85Ω for Rr. When the 180 degree dipole Rr of 85Ω is multiplied by 0.75 for a 120 degree V dipole, Rr = 63.75Ω, which is not in agreement with the software calculated Rr = 73.66Ω. The reason, as you will soon see, is that the entire inverted V dipole, unlike the 180 degree linear dipole, is not 30ft above ground.

The 120 degree V has an average height above ground, which is not the apex height. The *average* height of the 30ft high apex, 120 degree V dipole, is closer to 26ft (7.93m) above ground, about 0.37 wavelength above ground on 20M.

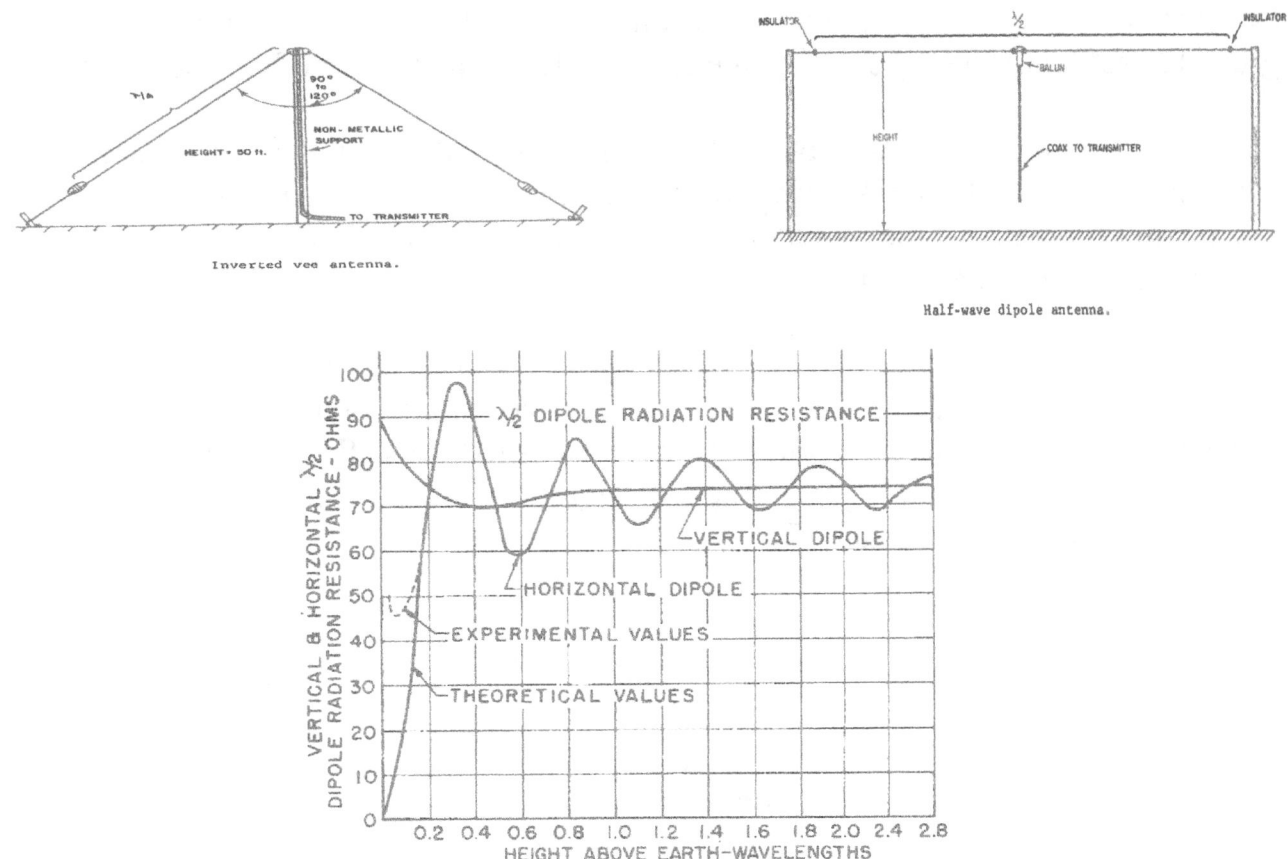

Inverted vee antenna.

Half-wave dipole antenna.

Looking at the graph for 0.37 wave above ground, I see around 97Ω. Multiplying 97Ω by 0.75 for the 120 degree V gives Rr = 73Ω, which is much more in agreement with the software Rr = 73.66Ω calculation.

For inverted V dipoles the actual height above ground is closer to the average height. This would require that an inverted V apex height be greater than a similar linear dipole height above ground for a realistic comparison.

The 120 degree inverted V antenna is still, mostly, a horizontally polarized dipole. There will be some vertical component. The ground image creates a parasitic element as it did for the linear dipole. The majority of the radiated energy at low heights will be up overhead due to the ground image.

An Application of the V Form to Elevated Monopole Radials

As a monopole with horizontal radials moves away from ground toward free space the radiation resistance, Rr, decreases. This can be seen at the end of the section "Reducing Counterpoise Length Using Inductive Loading" in which the antenna was raised above ground in software, clearly showing the Rr value reduced as the height is increased.

For a 1/4-wave monopole, with two 1/4-wave horizontal radials in free space, software calculated Rr = 23.89Ω at resonance, with four horizontal 1/4-wave radials, 22.54Ω. A two radial version had a measured radiation resistance of 21.0Ω in the 1940 article by the inventors, Brown & Epstein.

That the radiation resistance of a 1/4-wave monopole with horizontal 1/4-wave radials reduces toward around 23Ω when raised to free space, seems not to be presented in the usual antenna books, but is important. This is why radials of VHF and UHF verticals often droop to form an angle greater than 90 degrees from the monopole in order to raise the radiation resistance to nearer 50Ω.

Think of those drooping radials and the monopole as an inverted V dipole rotated 90 degrees. If the vertical dipole were completely straight the free space radiation resistance would be around 73Ω. By making the angle more than 90 degrees but less than 180 degrees a 50Ω radiation resistance can be had.

5.4 The Harmonically Operated Dipole

Thus far we have been adjusting the length of an antenna to resonance at a particular frequency. Through all the examples the aim was to treat the antenna as a transmission line and by using transmission line theory calculate lumped AC circuit component values. This method produced the equivalent of a series resonant circuit by treating the antenna as an open end transmission line. Like a transmission line, an antenna can have multiple resonances (LC circuits can't).

There is another view that will be useful in considering a *fixed length* transmission line, such as an antenna. Varying frequency, not length.

For a fixed length transmission line, or antenna, when the monopole or dipole is at the first resonance we have a1/4-wave ransmission line.

If we lower the applied frequency, the wave becomes too long to fit the 1/4-wave line, the line is shorter than 1/4-wave and the open transmission line input reactance will be capacitive. If we raise the applied frequency above 1/4-wave, the wave becomes too short to fit the fixed length 1/4-wave line. The open line is longer than 1/4-wave and the input reactance will be inductive.

If we continue increasing the frequency, going still higher, the antenna becomes a 1/2-wave, an equivalent parallel resonant circuit, with a high impedance.

When you continue increasing the frequency beyond the 2nd harmonic, to the 3rd harmonic, the electrical line length becomes 3/4-wave and you return to a series resonant circuit equivalent.

As we increased our frequency from 1/4-wave resonance to 1/2-wave anti-resonance the frequency doubled.

If we continue raising the frequency applied to the antenna it will have another resonance when the frequency is 3 times the 1/4-wave resonant frequency.

A common term for whole number multiples of a starting or base frequency is a harmonic. Twice a certain frequency is called a second harmonic. Three times, the third harmonic, etc. At the 2nd harmonic, a monopole or dipole will behave like a high impedance parallel resonant circuit, just like an open 1/2-wave transmission line will.

Due to this high impedance at the 2nd harmonic these antennas are called "voltage fed." The feed point is at the high voltage, low current location. And so it will be for an antenna (monopole or dipole), as the frequency doubles from resonance, the antenna becomes a 1/2-wave line and behaves as a high impedance parallel resonant circuit.

The original starting frequency goes by two names, fundamental and first harmonic. The term fundamental is the more common.

At odd harmonics (3,5,7...) the monopole or dipole will behave as a series resonant circuit and have a low impedance (current feed). At even harmonics (2,4,6....) the feed point impedance will have a parallel resonant circuit equivalence and a high impedance (voltage feed).

The RCA graph shows how the dipole antenna goes from odd frequency series resonant circuits (bottom dashed line) to high impedance parallel resonant circuits (upper dashed line) at even frequencies.

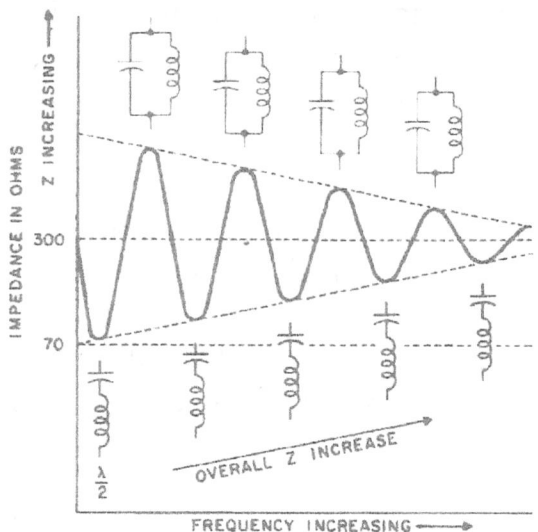

Figure 7-4. Dipole Impedance Excursions

Also notice from this graph all the series resonant odd harmonic frequency radiation resistances are greater than the 73Ω fundamental radiation resistance (the lower sloping dashed line).

Consider the HF amateur radio bands. There are both even and odd harmonic relationships between some bands. This then leads to the idea of antennas operated on harmonics.

A common HF example of antenna odd harmonic operation is the 40M monopole or dipole operated on 15M. Since 21MHz is three times (3/4-wave) 7MHz the antenna should exhibit series low impedance on both frequencies.

Does this mean a 40M antenna will be resonant at three times 7.1MHz, 21.3MHz?

Well, not really.

When we use the formula 234/f for a monopole or 468/f for a dipole, the resulting length has been modified. The physical length from these formulas are about 95% of the free space wave length for a conductor. As the radiator conductor circumference increases the needed per cent shortening becomes greater, as shown by the graph at the top of the next page.

91

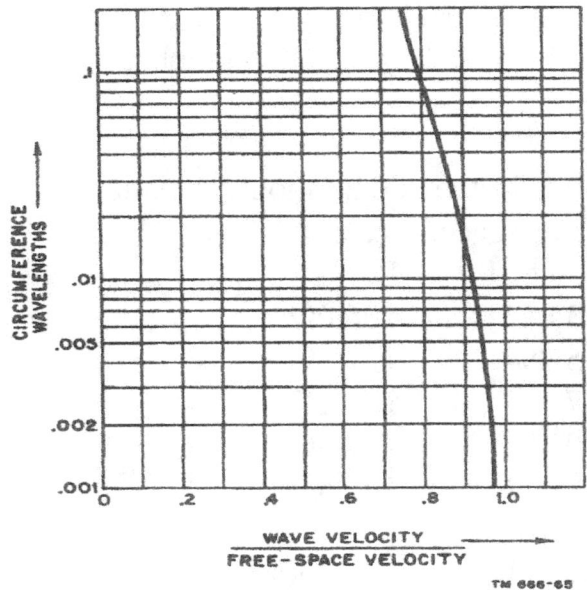

Figure 61. Effect of antenna circumference on wave velocity.

While the correction is necessary at the fundamental, it is not needed at the antenna harmonic frequencies. What this means is a monopole or dipole carefully adjusted for resonance at 7.1MHz will not have 3rd harmonic resonance at 21.3MHz.

The harmonic length does not require the shortening used for the fundamental. When the antenna is operated on a harmonic the 5% shortening does not apply to the harmonic as it does to the fundamental. The actual harmonic frequency of resonance will be higher than 21.3MHz because the antenna at its fundamental is short for the harmonics. When operated on an odd harmonic the antenna actual physical length should be closer to the free space length.

Here is what software calculated for a 7.1MHz wire dipole in free space:

7.1MHz	$72.03-j0.527\Omega$	1.44 SWR	
21.3MHz	$97.31-j71.14\Omega$	3.21 SWR	
21.643MHz	$106.0+j0.0506\Omega$	2.12 SWR	(3/4-wave resonance)

As far as software is concerned the fundamental frequency length reduction gives a free space physical length that is short on the 3rd harmonic indicated by the $-j71.14\Omega$ capacitive reactance at 21.3MHz.

Software found the actual harmonic resonant frequency to be 21.643MHz, not three times 7.1MHz, but, 3.048 times. At 21.643MHz the dipole is at 3/4-wave. The 40M antenna can still be used on 15M but it will have a higher SWR than at the fundamental frequency of 7.1MHz because of its reactance and higher radiation resistance.

The radiation resistance at the actual harmonic resonance is greater than at the fundamental resonance, as indicated by the RCA graph, and for 21.643MHz is 106Ω in free space by software.

Here is what software found for the odd harmonic radiation resistances and actual multiplier for a dipole in free space:

Harmonic	Actual	Rr
3rd	3.045	106.0Ω
5th	5.094	122.3Ω
7th	7.143	133.2Ω
9th	9.192	141.3Ω
11th	11.24	147.8Ω

The radiation resistance is higher at the harmonic frequencies than the fundamental, this is always the case for antennas operated on odd harmonics.

1) The actual harmonic frequency of operation will be higher than a simple multiple. This is because the formulas for the fundamental frequency length produces a shorter than free space antenna, which makes it short at all odd harmonics.
2) The odd harmonic frequency will be closer to the free space length.
3) The radiation resistance of the antenna at any odd harmonic resonance will be higher than the radiation resistance at the fundamental.

There is another effect associated with operating an antenna on its harmonics. The radiated field changes due to the additional length. The images below show what the harmonic fields look like compared to the fundamental 1/2-wave dipole in free space.

At the harmonic frequencies the antenna develops multiple lobes. This will be further discussed in the chapter "The Maximum Gain Dipole."

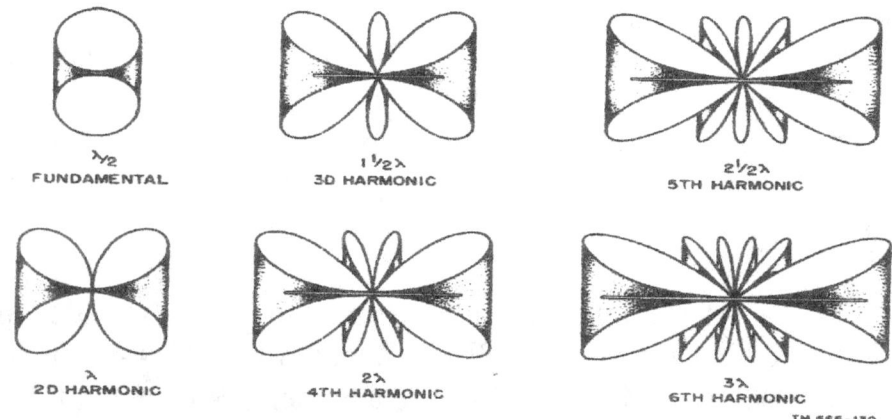

Figure 126.   Radiation patterns of harmonic antennas.

5.5 The Off-Center Fed Horizontal Dipole

The "Windom"

Some historical background on what Loren "Windy" Windom (8GZ/8ZG) saw in 1928 and 1929 at Ohio State University.

Windom was an active progressive Amateur Radio operator attending Ohio State studying Law. As might be expected he hung out with other students who were amateur radio operators, many were electrical engineering students.

At that time in radio a way to feed a horizontal 1/2-wave antenna was with a single-wire transmission line. A wire was attached to the antenna and came down vertically as a feeder. There was some disagreement as to the best location for the single wire feeder attachment point on the antenna and the single wire line characteristic impedance.

One experiment by the EE students (some held amateur radio licenses as well as their professor) was to find the feed point location for the single wire feeder to produce maximum transfer of power between the single wire vertical feeder and the horizontal antenna.

The resistance of a resonant horizontal 1/2-wave dipole should vary from low at the center, the current maximum, to high resistance at the open ends.

An experiment was devised to create a trolley like arrangement to investigate current in the single-wire feeder and the horizontal antenna wire as the feeder location was moved. The idea was, when the impedance of the single-wire vertical transmission line hit the spot where its impedance was the same as the feed point resistance of the horizontal antenna there would be maximum transfer of power indicated by the RF current meters.

Keep in mind that this off-center fed dipole was a consequence of a match for a particular diameter single-wire vertical transmission line connected to a horizontal dipole antenna of a particular diameter and height above ground.

The results of the experiments at Ohio State were published in The Proceedings of the Institute of Radio Engineers, October, 1929. Before that publication was issued, Windom wrote an article published in QST, September, 1929, entitled "Notes on Ethereal Adornments" about the experiments with the single-wire feeder. Thus, a reporter's name became an antenna type.

The single-wire feeder, for a time, was a popular way to feed a horizontal "short-wave" Hertz antenna. The single wire feeder, by being shifted from the middle of the dipole, also gave access to even harmonics at a lower radiation resistance than when the dipole was center fed.

The single wire feeder is also a specific length. This is in an effort to keep the feeder standing wave at a minimum so the majority of the radiation would be from the horizontal wire, in effect, attempting to keep the entire antenna from acting as a flat-top vertical monopole against earth.

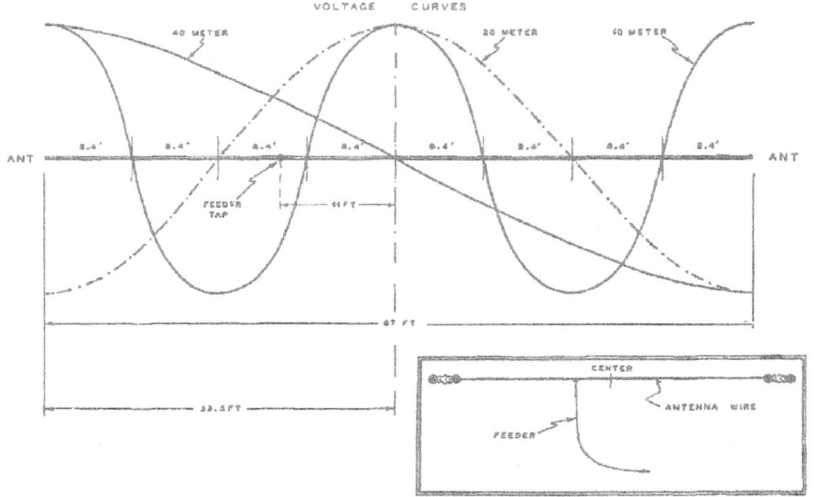

Figure 19. Single-wire-fed antenna for all-band operation. An antenna of this type for 40-, 20- and 10-meter operation would have a radiator 67 feet long, with the feeder tapped 11 feet off center. The feeder can be 33, 66 or 99 feet long. The same type of antenna for 80-, 40-, 20- and 10-meter operation would have a radiator 134 feet long, with the feeder tapped 22 feet off center. The feeder can be either 66 or 132 feet long. This system should be used only with those coupling methods which provide good harmonic suppression.

From Jones Antenna Handbook 1937

As the idea circulated over time, using a single-wire feeder fell out of favor for a parallel transmission line. While this may seem reasonable it had the effect of feeding an unbalanced antenna with a balanced transmission line.

Eventually the high impedance parallel transmission line was replaced by a transformer at the antenna feed point to reduce the off-set antenna impedance for a lower impedance balanced line and later an unbalanced coaxial feed line.

When the feed line is coax, the Windom antenna can return to a single-wire feed of the original antenna if the transformer at the feed point does not keep current from flowing on the outside of the coax shield. The 4:1 voltage baluns used at the feed point on this type Windom derivative will not prevent common mode current flow on the coax shield. A current choke/balun should be used to prevent current flow on the coax shield. (See "Don't Let Your Coax Ruin Your Antenna" in Afterwords and Afterthoughts in the Appendix.)

The radiation resistance of a resonant 1/2-wave dipole for any location along its length can be found from the relationship, $\mathbf{Rf = Rr/(cos^2\Theta)}$, where Rf is the feed point resistance, Rr is the radiation resistance at the current maximum location (center of the 1/2-wave dipole) and Θ is the distance in degrees from the center of the dipole (or the base of a monopole) out to the off-set feed point location. Zero degrees is at the current maximum and 90 degrees would be the current minimum, at the end.

Keep in mind that the Rr of a horizontal dipole above ground depends on its height above the ground image. We have seen in previous sections that dipole Rr varies with height and harmonic frequency.

Let's take the off-set feed formula for a test drive at three selected locations on a 1/2-wave dipole.

For 30 degrees out from the center the radiation resistance will be 1.3 times the center resistance. For 45 degrees the resistance will be twice Rr. For 60 degrees from the center the resistance will be four times Rr. The 60 degree location (or 30 degrees, 1/3, from the end) is a commonly suggested location for off-center feed in recent times. If you have a dipole in free space, four times 73Ω would be 292Ω, which would seem a good match for a 300Ω parallel feed line. Or with a proper 4:1 current balun, a 75Ω line.

This relationship of $Rr/(cos^2\Theta)$ begins to become inaccurate the closer you get to the end of the antenna at 90 degrees. If you take the cosine of 90 the calculator reports a zero, and anything divided by zero implies infinite resistance at the dipole end. Infinity may be a concept but it is not a number.

All antennas (dipole or monopole) have a real resistance at their ends, not infinite resistance. The open end resistance is dependent on the antenna Za and Rr, the resistance at the current maximum. Not only does height above ground effect the center radiation resistance, Rr, and Za, both have an effect on the actual resistance at the antenna open end and any off-center feed point (Rf).

To find the actual end resistance we need to treat the antenna as a 1/4-wave transformer. The actual resistance at the end of a 1/2-wave dipole or 1/4-wave monopole (quarter wave line), from transmission line theory is, $\mathbf{R_{end} = (Za)^2/(Ro)}$, where Ro is the radiation resistance at the current maximum, the center of the dipole, and Za is the characteristic impedance of the antenna as a transmission line. The open end resistance for a monopole was calculated in section 3.7.

Examination of this relationship tells us that as the center radiation resistance (Rr) of a horizontal dipole antenna becomes lower the resistance at the ends becomes higher. This also means the voltage at the end will become higher. The thinner the antenna conductor, the larger Za, the higher the end resistance and end voltage.

Short, thin, antennas, will have very high voltages at the end. A fluorescent tube, or neon bulb, held near the open end of an antenna will glow when RF is applied due to the high voltage at the end of the antenna.

As you learned in "The Horizontal Dipole Meets Ground" section, if the horizontal dipole is not in free space, the center radiation resistance (Rr) will be changed depending upon height above ground. And, so will any off-center fed radiation resistance. One of the outcomes of off-center feed is that the off-set feed point can be somewhere between Rr at the center and something very high at the end for the fundamental, and the even or odd harmonics. This is the basis of the Windom being a "multi-band" antenna.

All dipoles are multi-band, it's just that the center-fed 1/2-wave dipole varies from odd harmonic low radiation resistance to even harmonic high radiation resistance as shown in the previous chapter on harmonic operation. Off-center feed can provide intermediate radiation resistances, depending on the off-set feed point location and harmonic.

The $1/(\cos^2\Theta)$ relationship applies to the total impedance, so the Windom suffers from the same problems as the center fed 40M dipole when operated on 15M. If the antenna is resonant on the fundamental, the harmonics will not be whole number multipliers because the antenna is shorter than it should be on all harmonic frequencies, whether even or odd, free space or near earth.

Software bares this out. I constructed a virtual wire Windom dipole antenna in software and analyzed it in free space. I adjusted the virtual length for lowest SWR at 7.1MHz, and fed it 60 degrees from the center, a commonly recommended location for a modern Windom.

Here is what software calculated in free space:

Fundamental	7.1MHz	292.0-j0.106Ω.
Second	14.2MHz	115.2-j24.3Ω.
Third	21.3MHz	97.63-j39.13Ω.
Fourth	28.4MHz	147.4-j80.01Ω.

Our $Rf = Rr/(\cos^2\Theta)$ was validated for the fundamental, the software agrees with four times 73Ω for free space.

As was the case of the center fed 7.1MHz dipole when used on 21.3MHz, we see, by the -jX values, the antenna is short on all the harmonic frequencies, which indicates that the actual resonant frequencies are higher than a simple harmonic multiplier. Also, we see variations in the relocated feed point resistance at the harmonics compared to the fundamental.

Now let software bring the Windom down to earth. I used a height of 30ft (9.14m) above ground:

Fundamental	7.1MHz	355.9+j144.4Ω.
Second	14.2MHz	141.4-j38.63Ω.
Third	21.3MHz	95.06-j46.79Ω.
Fourth	28.4MHz	147.0-j88.52Ω.

We clearly see what was previously learned about center fed dipoles over real ground, the ground image changes resistance and reactance.

Why did the radiation resistance of the Windom change so much on 40M? Remember, shifting the feed point creates a multiplying effect. The center radiation resistance in free space was changed when the antenna came near ground, and the antenna is closest to ground on 40M. Check out the dipole radiation resistance vs height graph. The radiation resistance at the center of the dipole was reduced and multiplied by four.

Also notice the large change in reactance at 7.1MHz as the Windom is nearest to the ground image. The multiplying effect of off-set feed not only applies to the center radiation resistance, it also multiplies any center reactance too. The entire center impedance is multiplied.

The difference in free space impedance and real ground impedance at the higher frequencies (greater heights in wavelengths above ground) has less variation. The dipole is further away from its image at the higher harmonic frequencies. Remember that once a height of 1/2-wave or greater is reached, the Rr variation is less and closer to that of the free space value.

I did in software what might be done with the physical antenna, I adjusted the length of the virtual Windom for resonance at 7.1MHz, while 30ft above ground.

Software calculated:

Fundamental	7.1MHz	269.1-j0.529Ω.
Second	14.2MHz	117.0-j111.9Ω.
Third	21.3MHz	88.8-j143.0Ω.
Fourth	28.4MHz	184.1-j302.1Ω.

By adjusting the antenna for 7.1MHz resonance near ground the off-set feed point resistance returns to resonance with no reactance, but the capacitive reactance values of all harmonics have changed, making those values larger.

When operating antennas on harmonics, anything you do to optimize the antenna at any selected frequency, whether fundamental or harmonic, effects all the other frequencies, center or off-set fed. Over the years, in amateur radio literature, the off-set feed point location of the Windom has been moved around in an effort to optimize it for multi-band operation. Others have added reactance to the antenna at certain points, all in an effort to improve multi-band operation.

All software models of the Windom place the driving source at the feed point and there is no transmission line in the model. The real Windom will be fed, either with coax and a balun or a parallel wire line, not included here. So why not model it with a balanced feeder? Each circumstance will be different according to antenna height, the length of the line and whether it leaves the antenna at 90 degrees to the unbalanced Windom dipole. Again, the software knows nothing of the real environment or feed system.

For frequencies lower than 40M, over real ground, even the off center fed resistances will be very low, as we saw with the center fed dipole. For 160M, 80M (and perhaps 40M) it would be better to take a page from history and feed the Windom dipole as a flat-top against a counterpoise using a tuner. Connect the feeder from the Windom to the tuner and use a counterpoise as a capacitive ground. The more capacitance in your counterpoise, the better.

Here is an example of the feed method and impedance for a single wire off center fed antenna from U.S. Army TM 11-666 "Antennas and Radio Propagation" 1953, indicating 500 to 600Ω at the attachment point. Notice the 500 to 600Ω tap on the coil is above the ground tap, which tells us the impedance of the single wire feeder is with respect to ground. In the IRE paper of 1929 the authors wrote that the Zo of a single-wire transmission line ranges between 600 and 800 Ohms.

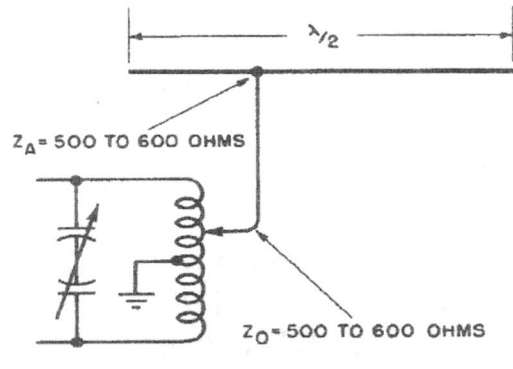

If we apply the formula for the characteristic impedance (Za) of a vertical wire to the suggested lengths of 33ft, 66ft, and 99ft, using #12 wire (from the IRE paper) 0.0808 inch or 2.0525mm diameter, the calculated Za will be the Zo of the single wire feeder, which are 491Ω, 533Ω, and 557Ω respectively. Treating the 33ft, 66ft, and 99ft feed wires as a monopole over perfect ground, at half resonance (1/8-wave), software found Zo of 503Ω, 545Ω and 570Ω respectively. These values more closely agree with the U. S. Army's 500 to 600Ω than the original 1929 estimate of 600 to 800 Ohms.

We can apply the Za formula for a monople antenna to the single wire feeder because the original single wire fed Windom *relies* on an earth image.

The Army manual does not show, exactly, where the single wire feeder should be attached to the dipole. The best spot for maximum transfer of power would be a point with an off-set resistance of 500 to 600Ω. That off-set location will vary with the dipole height above ground, as software modeling has shown.

The "W3EDP"

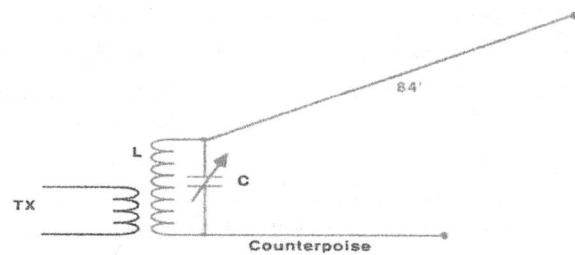

The W3EDP off-center fed antenna was first reported in QST, March 1936. Different lengths of counterpoise wires were used (and suggested). I have used this type OCF antenna from a second floor room with an antenna tuner replacing the tuned circuit. The antenna and counterpoise were connected directly to the antenna tuner.

5.6 An Inductively Loaded Horizontal Dipole

The goal in this example will be to use inductive loading of a 30M band length wire dipole for operation on 40 meters.

The length of the wire dipole for 10.12MHz is (468/10.12MHz) = 46.3ft (14.1m). The calculated average characteristic impedance of this antenna as a transmission line will be Zo = 120[Ln(2)(23.2ft)(12in/ft)/(0.0641in) - 0.75] = 998Ω (software calculated Zo = 999.9Ω using half the resonant frequency method).

The electrical length of the 30M dipole at 7.1MHz will be, (7.1MHz/10.12MHz)(90 degrees) = 63.14 degrees. The feed point capacitive reactance of the 30M dipole at 7.1MHz will be, (998Ω)/(tan 63.14 degrees) = -j505Ω (software calculated – j505.8Ω).

The capacitive reactance can be canceled at the feed point with an inductive reactance of +j505Ω. The required inductance at 7.1MHz will be (505Ω)/(2π)(7.1) = 11.3µH. With 11.3µH of inductance at the 30M dipole feed point, at 7.1MHz, software calculated Z = 27.04+j1.778Ω, 1.85 SWR.

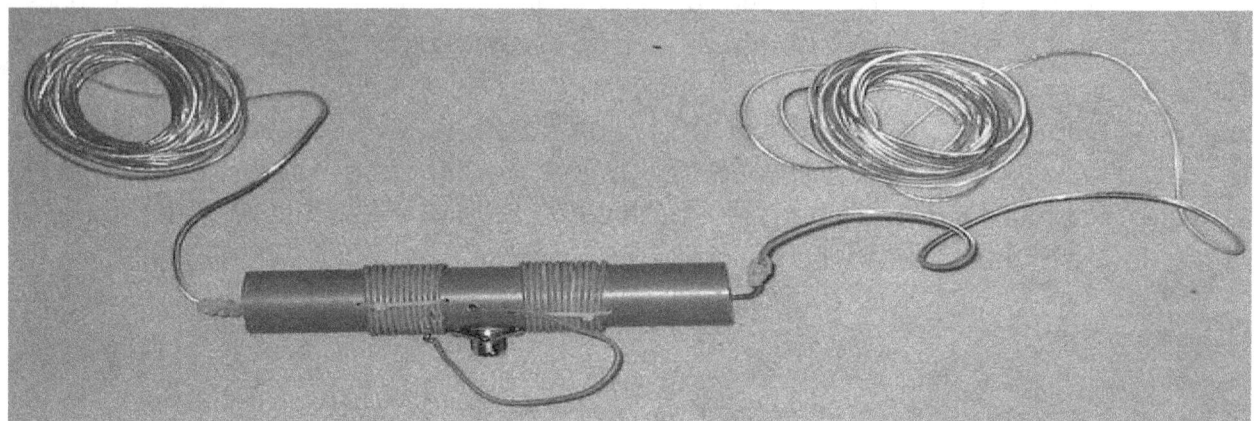

An example of a home brew, feed point inductively loaded, short dipole.

In the center loaded monopole chapter we discovered a short cut for finding the reactance of the inductor for center loading the monopole to be twice the reactance at the feed point. It worked for loading a counterpoise too, let's try it on a dipole.

Let's load the center of the poles of the 30M dipole for 40M instead of the feed point and use our short cut. The capacitive reactance of the dipole at the feed point was calculated to be -j505Ω using the antenna as transmission line model. Keeping in mind that a dipole is two series connected monopoles at their feed points, this means each half (pole) of the dipole is responsible for half the total -j505Ω center reactance.

Using our doubling short cut from the monopole example, we should double half the dipole center capacitive reactance, which brings us back to -j505Ω. Each pole of the dipole center loading coil will need an inductive reactance equal to the entire center feed point capacitive reactance.

Perhaps we now have a short cut for center loading dipole halves? Let's see.

Using our short cut for center loading each half of the dipole will require +j505Ω of inductive reactance in the center of the poles of the dipole for 7.1MHz, which has already been calculated to be 11.3µH.

Placing 11.3μH inductance at the centers of both poles of the 30M dipole, at 7.1MHz, software calculated Z = 45.08+j24.38Ω, 1.68 SWR.

It appears we do have a trick for center loading the poles of a dipole. The inductive reactance of the loading coils should be equal to the total capacitive reactance at the dipole feed point in the center of each side.

Notice, as was also the case for inductive loading of the monopole, placing the loading coil at the center of the poles, rather than the feed point, nearly doubles the radiation resistance. According to software, the radiation resistance went from 26.0Ω to 45.08Ω, not only reducing SWR but improving antenna efficiency.

Loading inductors placed in the two poles of a dipole.

A Two-Frequency Inductively Loaded Dipole?

In April 1961 QST published an article by William Lattin, W4JRW, entitled "Multiband Antennas Using Loading Coils."

His idea was, at a higher operating frequency, the inner dipole would have its ends terminated with inductors. Wire was added beyond the inductors for a lower frequency inductively loaded antenna.

The general idea is that the reactance of the loading coils at the higher frequency would be larger than at the lower frequency and act as RF "chokes" so little to no current would flow into the extensions beyond those loading coils at the higher frequency.

Let's investigate this idea using software to model the 30M length antenna, just designed, center loaded in both poles for operation on 40M.

A free space wire dipole model was constructed in software, giving 72.07 -j0.525Ω at 10.12MHz.

With the coils being placed in the center of the dipole poles, the inner dipole (which would be the higher frequency dipole) will have a length one half the total length and be, in theory, resonant around (468/23.3ft) = 20.2MHz.

Since the 11.3µH loading coils were selected for 7.1MHz, the reactance of the coils at 20.2MHz will be +j1433Ω. Will 1434Ω of inductive reactance at the end of the center, higher frequency, dipole be enough to approximate an open at 20.2MHz?

For the loaded 30M dipole on 20.2MHz software calculated, $Z = 18.4-j303.7\Omega$ at the feed point. A sweep by the software found resonance at 23.631MHz, with an impedance of $Z = 50.59+j0.0747\Omega$, 1.01:1 SWR.

What's going on here? Let's figure this out by antenna as transmission line logic.

The inner dipole is acting as a 1/4-wave transformer, which inverts the net inductive reactance at its end to a capacitive reactance ($-j303.7\Omega$), at the other end, the dipole feed point. Put an inductive reactance at the end of a 1/4-wave transmission line and it will be inverted to a capacitive reactance at the other end, including antennas.

Adding series capacitance at the feed point of an antenna will electrically shorten the antenna, just the same as a physical capacitor, and make the actual resonant frequency become higher. The net inductive reactance, and therefore the feed point capacitance transformed, is dependent on the net inductance at the end of the inner dipole and Za.

A change in the length of the wire beyond the inductor causes a change in the resonant frequency of the inner dipole too, because it changes the net inductive reactance at the end of the inner 1/4-wave transformer, the inner dipole.

In his article Lattin used 120µH for an 80 and 40M dipole. At 7.1MHz the reactance of 120µH would be 5353Ω. He wrote "We have not found any exact formulas to determine the relationship between the lengths of wire, loading coils, and the two frequencies."

As my notes clearly show (but I did not understand why at that time) from experiments years ago, this type antenna required quite a bit of adjusting for success and did not always produce a two frequency antenna on the frequencies I had hoped for.

Next I will show how a particular type "trap" (parallel resonant circuit) can more easily make a two-frequency antenna.

5.7 A Two Frequency Antenna Using a "Trap"

The word trap, as used in radio history comes from wave trap or frequency trap. Wave traps were parallel resonant circuits, usually used to stop or attenuate to a certain level an unwanted signal on the antenna at the receiver input.

The traditional method of making a two frequency trap antenna is for the trap to be parallel resonant at the higher of the two desired frequencies.

A parallel resonant circuit, at resonance, will have a high impedance. This high impedance is then placed at the distant end of either a 1/4-wave monopole or in the poles of a 1/2-wave dipole whose resonant length is the higher of the two frequencies of interest. The high impedance of the trap isolates the higher frequency inner length from any conductors that are placed beyond the trap.

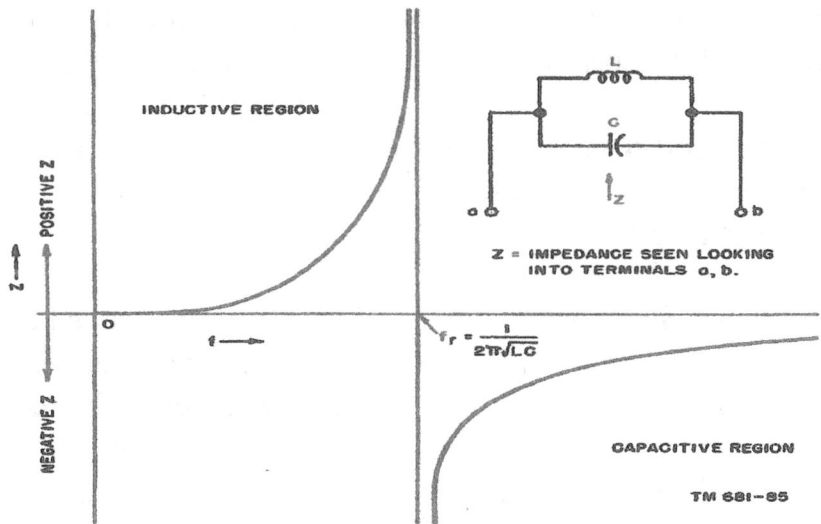

Figure 101. Reactance chart for ideal parallel tank circuit.

At the lower frequency, the conductor beyond the trap, with the net inductive reactance of the trap and the inner length, creates an antenna that can be resonant at a lower frequency, at which point the trap effectively becomes a loading coil.

One of the challenges of this traditional parallel resonant trap design is the unlimited number of choices of inductors and capacitors that can be used to form the high impedance trap.

Also this method creates a high voltage across the trap at resonance. And two traps are required for a dipole antenna, one in each poleWhatever the choice, the net equivalent inductance of the trap, at the lower frequency, is very much dependent on the inductor and capacitor used to form the trap. All these variations have an effect on the final length of the wire beyond the trap to cause resonance at the lower frequency.

These variations create a combination of moving parts that results in only one sure thing, the length of the inner higher frequency monopole or dipole making up the antenna. And, that the trap will be located somewhere along the length of the radiator.

There is another approach for a parallel resonant circuit that provides a two frequency antenna using a circuit that is *not* resonant at the higher of the two frequencies that I have developed. Taking this approach will eliminate the moving parts challenge.

Before getting into the specifics of this different design method, let's review some facts about antennas that are too short, or too long, at a particular frequency.

An antenna that is too short for the desired frequency will have a capacitive reactance at its feed point. Alternately, an antenna longer than the desired operating frequency will have an inductive reactance at its feed point.

An antenna that is physically short for a particular frequency can be brought to resonance through the use of a loading coil to cancel the capacitive reactance. The loading coil has been placed at the feed point, or at the center of the radiator for short antennas in previous examples. We have also seen that an antenna too long for a higher frequency can be loaded with a capacitor to cancel its inductive reactance. As with the inductive load for the short antenna, the capacitor for the longer antenna can be placed at the feed point, or the center of the radiator.

The secret to my two frequency antenna, presented here, is for the basic antenna to be resonant at the *geometric mean* frequency of the two frequencies of interest.

The geometric mean frequency is the square root of the product of the high frequency and the low frequency of the two frequency antenna. Mathematically, $\mathbf{Fg = \sqrt{(F1)(F2)}}$. (A geometric mean related to transmission lines was used in the section "The 1/4-Wave Monopole – An Impedance Transformer")

By placing a properly selected inductance and capacitance as a parallel circuit resonant on the geometric mean frequency, Fg, at the feed point of the radiator, or along the radiator, we can get the correct inductive reactance at the low frequency, and correct capacitive reactance at the high frequency needed for antenna dual resonance, automatically. The selection of the inductance and capacitance for the required geometric mean parallel circuit can be calculated, no guessing required.

A unique feature of the geometric mean method, as opposed to the traditional trap antenna, the geometric mean circuit can be placed at the feed point of the antenna. For a dipole this means only one parallel resonant circuit is required. With a traditional trap, two traps are required, one in each pole of the dipole. The traditional trap must be placed between the feed point and the end of the antenna, not at the feed point.

Why a geometric mean? Let's examine a very important formula associated with all resonant circuits, series or parallel. Resonance is calculated from the basic formula, $f = 1/(2\pi)(\sqrt{LC})$. Do you see it? A geometric mean in the resonance formula, the $(\sqrt{LC})$.

By making the basic antenna resonant at the geometric mean frequency of the two frequencies of interest we can take advantage of the geometric mean of L and C for a parallel resonant circuit. This is done using a parallel circuit resonant at the geometric mean frequency and placed at the feed point or along the geometric mean antenna, dipole or monopole. The Q of the series resonant circuit that is the antenna can nail down the L and C values required for the trap. All the moving parts have stopped moving and no random selection of L or C is required.

As an example, the geometric mean of 14.1MHz and 18.1MHz, $Fg = \sqrt{(F1)(F2)} = \sqrt{(14.1)(18.1)} = 15.98$MHz. Our dipole or monopole will need to be resonant at this frequency. Since the 15.98MHz antenna is short at 14.1MHz, it will have a capacitive reactance. At 18.1MHz the antenna will be long and have an inductive reactance.

Getting from a series circuit antenna to a parallel resonant circuit.

To calculate the parallel circuit L and C and make the antenna operate on 14.1MHz or 18.1MHz automatically, we need to do two things.

First, calculate the geometric mean of the 14.1MHz -jX and the 18.1MHz +jX. Ignoring the signs, $\mathbf{Xg = \sqrt{(+jX)(-jX)}}$ and find the geometric mean of the reactance at the two frequencies of 14.1 and 18.1MHz.

Second, find the Q of the antenna. $\mathbf{Q = Fo/BW} = Fo/(F1-F2) = 15.98$MHz/(18.1MHz - 14.1MHz) = 3.995 for this example. Since we are constructing a parallel resonant circuit we will be using the reciprocal of Q to calculate the L and C of the parallel circuit.

The reactance of the L and the C, which must be equal at the resonance frequency of 15.98MHz, is $(1/3.995)(Xg) = XL = XC$, which at 15.98MHz will be the reactance needed for the inductor and capacitor. In a resonant circuit, $XL = XC$. Inductance $L = XL/(2\pi)(Fo)$ and capacitance $C = 1/(2\pi)(Fo)(XC)$, at 15.98MHz for this example.

First I'll do a conduit monopole over perfect ground example for simplicity, then I'll do a wire dipole example.

The required length of the 1/4-wave monopole will be (234)/(15.98MHz) = 14.64ft (4.46m). The average characteristic impedance of the conduit monopole as a transmission line will be, $Z_a = 60[Ln(2)(12in/ft)(14.64ft)/(0.922in) - 1] = 297\Omega$.

The electrical length of the conduit monopole at 14.1MHz will be, (14.1MHz/15.98MHz)(90 degrees) = 79.41 degrees. The capacitive reactance at the feed point will be $-jX = (297\Omega)/(\tan 79.41 \text{ degrees}) = -j55.5\Omega$. The electrical length of the conduit monopole at 18.1MHz will be, (18.1MHz/15.98MHz)(90 degrees) = 101.9 degrees. The inductive reactance at the feed point will be $+jX = (297\Omega)(\tan 101.9 \text{ degrees} - 90 \text{ degrees}) = (297\Omega)(\tan 11.9 \text{ degrees}) = +j62.2\Omega$.

Now that we have the two reactance values at the feed point for 14.1MHz and 18.1MHz, calculate the geometric mean reactance (ignore signs) which is $X_g = \sqrt{(55.5\Omega)(62.2\Omega)} = 58.8\Omega$. The required reactance of the feed point parallel resonant circuit will be equal to $(58.8\Omega)/(Q) = (58.8\Omega)/(3.995) = 14.7\Omega$ for the L and C at 15.98MHz. The inductance L = 0.150µH and C = 677pF.

Using the LC values calculated by the transmission line model, here is what the computer model calculated:

14.1MHz   Z = 23.7+j4.919Ω   2.14 SWR.
18.1MHz   Z = 56.38-j11.5Ω   1.28 SWR.

The geometric mean circuit can also be placed along the radiator, rather than at the feed point, if you wish. Applying our short cut for center loading a monopole, we simply double the base reactance values. The needed center loading inductive reactance for 14.1MHz will be $(2)(+j55.5\Omega) = +j111\Omega$ and the needed center capacitive reactance for 18.1MHz will be $(2)(-j62.2\Omega) = -j124\Omega$.

The geometric mean of these two center loading reactance values will be $X_g = 117\Omega$. The trap required inductive and capacitive reactance will be $(117\Omega)/(Q) = (117\Omega)/(3.995) = 29.3\Omega$, twice what was calculated for the base. The the parallel resonant circuit for the center of the monopole will require L= 0.29µH (twice that for the feed point trap) and C = 340pF (half that required for the feed point parallel circuit).

Using these values, here is what software calculated for the parallel circuit at the center of the conduit monopole:

14.1MHz   Z = 30.44-j1.49Ω   1.65 SWR.
18.1MHz   Z = 41.01+j0.8525Ω   1.22 SWR.

With the geometric mean method, the LC circuit at the two operating frequencies of 14.1MHz and 18.1MHz, unlike the traditional trap, is never at resonance and will not be subjected to the very high voltages or large circulating currents of resonance.

A geometric mean feed point dipole

The greatest difference between a traditional trap dipole antenna and a geometric mean dipole antenna, one parallel circuit can be placed at the feed point.

Now for a geometric mean circuit at the feed point of a 15.98MHz wire dipole.

The characteristic impedance of the wire dipole antenna as a transmission line will be, $Za = 120[Ln(2)(14.64ft)(12in/ft)/(0.0641in) -0.75] = 943\Omega$ The electrical length at 14.1 and 18.1MHz remain the same as for the monopole.

At 14.1MHz, $-jX = (943\Omega)/(\tan 79.41 \text{ degrees}) = -j176\Omega$. At 18.1MHz, $+j = (943\Omega)(\tan 11.9 \text{ degrees}) = +j199\Omega$. The geometric mean reactance of these two is, $Xg = 187\Omega$. The trap L and C reactances will be $(187\Omega)/(Q) = (187\Omega)/(3.995) = 46.8\Omega$. This results in $L = 0.466\mu H$ and $C = 213pF$.

Using the calculated L and C values for a parallel resonant circuit at the dipole feed point, the 15.98MHz wire dipole, up about 30ft (9.14m), software calculated,
14.1MHz $Z = 61.02+j18.34\Omega$ 1.47 SWR.
18.1MHz $Z = 95.61+j26.99\Omega$ 2.11 SWR.

Construction and measurement of a geometric mean vertical monopole with a two wire counterpoise

The geometric mean monopole over perfect ground, as always, is the most simple case to calculate. But, the perfect ground is theoretical.

I have shown in previous sections that a monopole with an electrical length greater than 90 degrees can give a higher radiation resistance than 36.6Ω and a feed point inductive reactance that can be canceled by a capacitive counterpoise.

In the section "Using Capacitive Bottom Loading – The Counterpoise" the inductive reactance of the 50Ω monopole was canceled with the capacitance of a 2-wire counterpoise.

The 50Ω monopole had an electrical length of 100.5 degrees. To calculate the physical length of a 100.5 degree monopole, (100.5 degrees/90 degrees)(234) = 261. The calculated physical length of the geometric mean monopole for this Rr = 50Ω monopole example will be, (261)/(15.98MHz) = 16.33ft, (4.98m).

Since the geometric mean monopole is longer than 90 degrees at 15.98MHz it will have an inductive reactance at the feed point which will be canceled with a capacitive counterpoise.

The average characteristic impedance of this conduit monopole is, Za = 60[Ln(2) (16.33ft)(12in/ft)/(0,922in) -1) = 303Ω.

The 15.98MHz inductive reactance at the monopole feed point will be +jX = (303Ω)(tan 100.5 degrees – 90 degrees) = +j 56.2Ω. Software calculated 52.74+j 57.32Ω over perfect ground (2% difference).

Here is what software found for the 100.5 degree 15.98MHz conduit monopole model over perfect ground:

14.1MHz	Z = 33.52-j10.55Ω.
15.98MHz	Z = 52.74+j57.32Ω.
18.1MHz	Z = 90.82+j142.0Ω.
14.33MHz	Z = 35.42-j2.395Ω, resonance.

Interestingly, the geometric mean of Rr = $\sqrt{(Rr1)(Rr2)}$, where Rr1 is the radiation resistance at the lower frequency and Rr2 is the radiation resistance at the higher frequency. Applying this geometric mean radiation resistance formula Rr = $\sqrt{(Rr1)(Rr2)}$ to the software values, Rr = $\sqrt{(33.52Ω)(90.82Ω)}$ = 55.2Ω, software, 52.74Ω

Now for the two-wire, co-linear, counterpoise. The counterpoise must cancel +j56.2Ω at the geometric mean frequency. To do that each wire must provide (2)(-j56.2Ω) = -j112Ω.

Using a wire 1ft above ground Zo = 60[Ln(4)(12)/(0.0641)] = 397Ω. At 15.98MHz the wavelength will be, (984)/(15.98MHz) = 61.6ft.

The required electrical length of the single wire open transmission line will be, Θ = tan⁻¹ (Zo)/(-jX) = tan⁻¹ (397Ω)/(112Ω) = 74.19 degrees. The physical length will be (74.19 degrees/360 degrees)(61.6ft) = 12.7ft (3.87m), each. At 15.98MHz software calculated Z = 48.05-j2.091Ω for a counterpoise wire length of 4.0m (13.1ft), which is 3.25% longer than the transmission line model of 3.87m.

The software model, with the 4m counterpoise wires:

14.1MHz   Z = 31.4-j96.99Ω.
15.98MHz   Z = 48.05-j2.091Ω   1.06 SWR.
18.1MHz   Z = 79.36+j109.3Ω.

The *measured* impedance with two 12.5ft (3.81m) counterpoise wires were:

14.1MHz   Z = 33.9-j98.6Ω   7.7 SWR.
15.98MHz   Z = 56.5-j12.7Ω   1.31 SWR
18.1MHz   Z = 86.7+j92.3Ω   4.0 SWR.

The geometric mean of the measured reactance at 14.1 & 18.1MHz
$X_g = \sqrt{(98.6Ω)(92.3Ω)} = 95.40Ω$. The required geometric mean reactance, 95.40Ω, must be divided by Q, which for this example remains 3.995, (95.40Ω/3.995) = 23.9Ω.

For a reactance of 23.9Ω at 15.98MHz, L = 0.238µH and C = 417pF, forming a parallel resonant circuit to be placed in series at the feed point.

Using my antenna analyzer I measured capacitors on hand and managed a 413pF total. Then I wound an air core inductor with #12 bare copper with more than the required inductance. Using the antenna analyzer I adjusted the inductor turns spacing, as close as possible, to 15.98MHz resonance with the 413pF capacitance.

Placing the tuned parallel circuit in series at the feed point the measured results:

14.1MHz   Z = 41.8+39.7Ω   2.4 SWR.
18.1MHz   Z = 95.4-j15.1Ω   1.97 SWR.

Here is the dual resonance plot of the geometric mean monopole with the two-wire counterpoise and the added geometric mean parallel resonant circuit at the feed point as measured by an antenna analyzer.

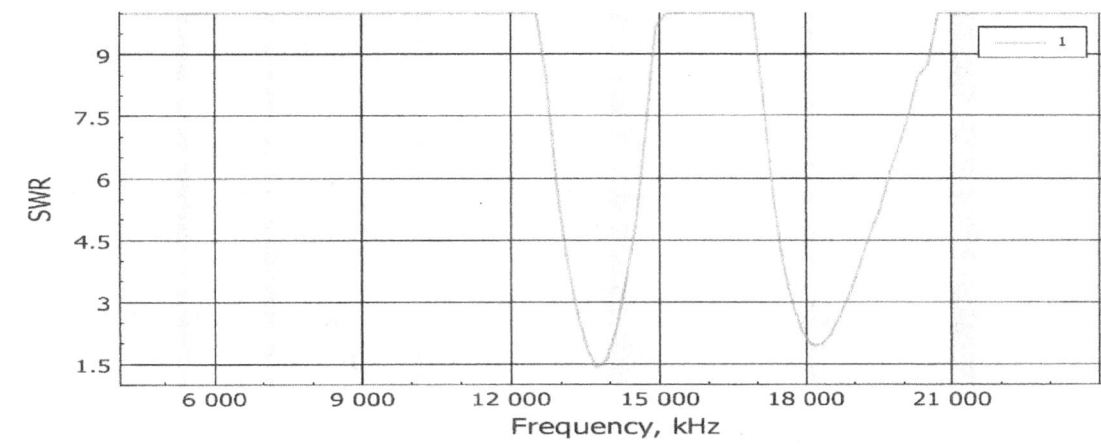

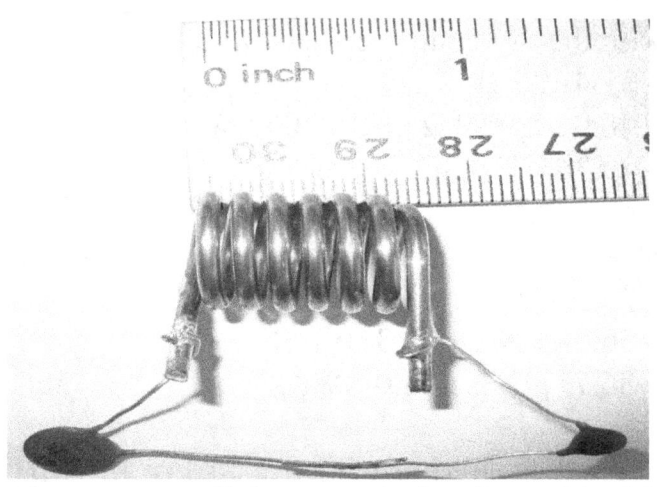

The geometric mean parallel resonant circuit used in this example.

The two measured minimum SWR frequencies are, 1.42 SWR at 13.700MHz and 1.97 SWR at 18.100MHz.

I've found that if you have the test equipment to measure the reactance of the antenna for the two frequencies and use those to find the L and C for the parallel resonant circuit, very good results can be had. When actual measurements are made you are seeing the antenna in its environment.

The how and why of the geometric mean parallel circuit.

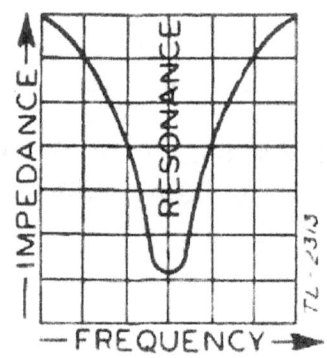 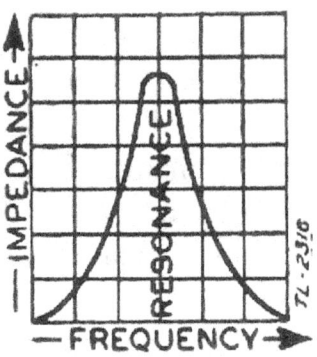

On the left is the impedance vs frequency graph of a series resonant circuit. The dipole behaves approximately like the series resonant circuit with low impedance at resonance.

On the right is the impedance vs frequency graph of a parallel resonant circuit. The geometric mean LC circuit will behave like a high impedance parallel resonant circuit.

The responses of the series and the parallel LC circuits are opposites as frequency varies. As the frequency is increased, the antenna becomes inductive, but, the parallel resonant circuit reactance becomes capacitive. For a decrease in frequency the series circuit antenna becomes capacitive but the parallel resonant circuit becomes inductive.

Since the circuit Q and the antenna Q are made close to the same, as well as their resonant (center) frequencies, when the series graph is superimposed on the parallel graph they intersect at F1 and F2. the dual frequencies of the geometric mean antenna.

The intersections of the series circuit antenna and parallel resonant circuit results in a cancellation of reactance at F1 and F2 leaving the feed point resistance (and resistive losses of the parallel tuned circuit, antenna and ground system) at those two frequencies.

As explained in the section "Antenna Reactance, Resistance, Bandwidth and "Q"", the Q of a series circuit is X/R. For a parallel circuit Q is R/X, the inverse of the series circuit. This makes Q the connection between a series AC circuit (the antenna) and a parallel AC circuit (the geometric mean parallel circuit). For every series RLC AC circuit there is an equivalent parallel RLC AC circuit and vice versa.

This is why some antenna analyzers can give a measured impedance as an equivalent series circuit, or as an equivalent parallel circuit.

5.8 The Bent Ends Horizontal Dipole

In this example I will show the steps to capacitively end load a 30M length horizontal dipole for operation on 40M, by bending the dipole ends. The bent ends form an open parallel wire transmission line providing capacitive end loading to the short dipole.

Looking at the dipole radiation resistance graph for a dipole at 30ft (9.14m) above ground, about 0.31 wavelengths on 30M, should place the radiation resistance around 90Ω.

Software calculated Z = 90.71-j22.73Ω for a basic wire dipole on 10.12MHz, 46.3ft (14.1m) in length, 0.31 wave above ground, with an SWR of 1.97. The 0.31 wavelength height at 10.12MHz is about 0.22 wave length above ground for 7.1MHz. From the graph, at 0.22 wave above ground, I see 70Ω.

Using a calculator, let's see what we have for Rr and -jX on 40M.

The 30M wire dipole characteristic impedance, Za = 120[Ln(2)(23.15ft)(12in/ft)/ (0.0641in) - 0.75] = 998Ω.

The electrical length of the antenna on 7.1MHz, (7.1MHz/10.12MHz)(90 degrees) = 63.14 degrees. The estimated radiation resistance, with no end loading $R_r = R_o[(1-\cos\Theta)^2/\sin^2\Theta]$, using Ro from the graph at 7.1MHz will be, $R_r = 70\Omega[(1-\cos$ 63.14 degrees)^2/(sin^2 63.14 degrees)] = 26.4Ω.

The capacitive reactance of the short antenna at the center feed point will be $-jX = (Z_a)/(\tan\Theta) = (998\Omega)/(\tan 63.14) = -j505\Omega$. For an estimated total of, Z = 26.4-j505Ω for the 30M wire dipole on 7.1MHz. Software calculated, Z = 26.07-j523.3Ω on 7.1MHz for the 30M dipole up 30ft.

On 40M the dipole is missing length. In electrical degrees the missing length is (90 degrees - 63.14 degrees) = 26.86 degrees. The missing capacitive reactance of this length can be found using basic transmission line theory, $-jX = (Z_a)/(\tan\Theta) = (998\Omega)/(\tan 26.86$ degrees) = -j1971Ω.

Another way to get this reactance value is to go out to the end of the dipole with your virtual eye, look down toward the feed point where you will see a short, which will produce an inductive reactance. To find this inductive reactance, using basic transmission line theory, $+jX = Z_a(\tan\Theta) = (998\Omega)(\tan 63.14$ degrees) = +j1971Ω which will require -j1971Ω at the ends to cancel the reactance at the feed point and cause resonance at 7.1MHz. Same result, different approach.

As stated, I intend to model the bent ends as an open wire parallel transmission line to provide capacitve end loading. Think of this antenna as two horizontal inverted "L" wire monopoles connected in series at their feed points.

The characteristic impedance of a parallel transmission line is, $Z_o = 120Ln(2)(S)/(d)$ where spacing "S" between the conductors and the diameter "d" are in the same units. The characteristic impedance of a parallel wire transmission line with a 46.3ft (14.1m) spacing, the end to end length of the 30M dipole, characteristic impedance, $Z_o = 120Ln(2)(46.3ft)(12in.ft)/(0.0641in) = 1171\Omega$.

I already have calculated that I need -j1971Ω of capacitive reactance added to end load the 30M dipole for resonance on 40M. Now that I know the Zo of the parallel transmission line being used to provide that capacitive reactance I can find the line length in degrees, then in feet, for end loading.

The line length in degrees, from basic transmission line theory, will be, $\Theta = \tan^{-1}(Z_o)/(-jX) = \tan^{-1}(1171\Omega)/(1971\Omega) = 30.72$ degrees. To find the physical length I need to calculate the length of a wave at 7.1MHz, in free space, which will be (984)/(7.1) = 138.6ft.

The physical length of the parallel wire transmission line bent ends will be (30.72 degrees/360 degrees)(138.6ft) = 11.8ft (3.60m) on each end.

Continuing with the scientific calculator I will now find the estimated radiation resistance for the end loaded 7.1MHz dipole. For an end loaded antenna $Rr = Ro(\sin\Theta)^2$. In this example $\Theta = 63.14$ degrees. For Ro, according to the graph, the radiation resistance at 0.22 wavelength above ground is 70Ω. Substituting $\Theta = 63.14$ degrees and $Ro = 70\Omega$, I estimate $Rr = 56\Omega$.

After adding the two parallel end wires to the dipole, 11.8ft each to the 30M wire dipole model, software calculated, $Z = 64.69+j54.13\Omega$, 2.59 SWR. Using my calculator I got $Rr = 56\Omega$, telling me the SWR should be closer to $(56\Omega)/(50\Omega) = 1.12$. This tells me the end loading wires need adjustment (as does the $+j54.13\Omega$ from software if I didn't have an antenna analyzer). Adjusting the length of the end loading wires to 10.7ft (3.26m) each, software calculated, $Z = 59.88-j0.2884$, 1.2 SWR. The actual dipole in its environment may be different.

As was the case for the inverted L monopole, the bent antenna was longer than 1/4-wave, and that is the case for the bent ends 1/2-wave dipole. The length of a 40M 1/2-wave dipole would be, 468/7.1 = 65.9ft (20.1m). The total length of the 30M dipole with the bent end loading is, 46.2ft + 2(11ft) = 68.2ft (20.8m) total. As with the inverted L, the transmission line end loaded dipole is not just a bent 1/2-wave dipole.

The characteristic impedance of the end loading transmission line is not the same as the characteristic impedance of the antenna as a transmission line. A bent end antenna, whether monopole or dipole, will need to be longer than the calculated 1/4 or 1/2-wave length.

I modeled this antenna with the bent ends hanging down. If in an actual installation that is unsafe or not possible, remember those ends form a parallel transmission line, keep them as parallel as possible.

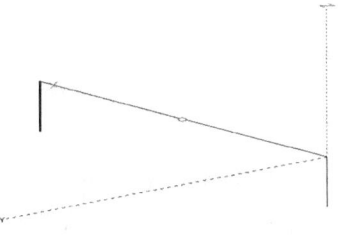

In the examples for shortening monopoles using end loading, a graph of the current distribution showed the majority of the current was in the first 60 degrees beyond the current maximum at the feed point. This left the end 30 degrees as a good choice for replacement with some type of capacitive end loading. In this example 30.72 degrees was replaced with capacitive end loading wires to resonate the 30M dipole on 40M.

An end loaded 60 degree monopole or dipole will give a field strength very close to the full 90 degree antenna element. The end 30 degrees contributes very little to the total radiated field due to the small current near the end of the antenna.

Retaining the first 60 degrees of the 90 degree current waveform of an antenna has lead to a sort of 2/3 (60/90) rule for short monopoles and dipoles.

Unfortunately, at some point, the missing 1/3 part was just forgotten about and the 2/3 monopole or dipole became some sort of "special" antenna length. If the 2/3 length antenna is not end loaded it's just another non-resonant radiator and is certainly not special in any way.

An important take-away from this example, it's better to bend the ends of a dipole than to make a short inductive loaded dipole. The radiation resistance can be higher with capacitive end loading and no need to wind coils. Keep in mind the entire dipole must be longer than 1/2-wave if you bend the ends.

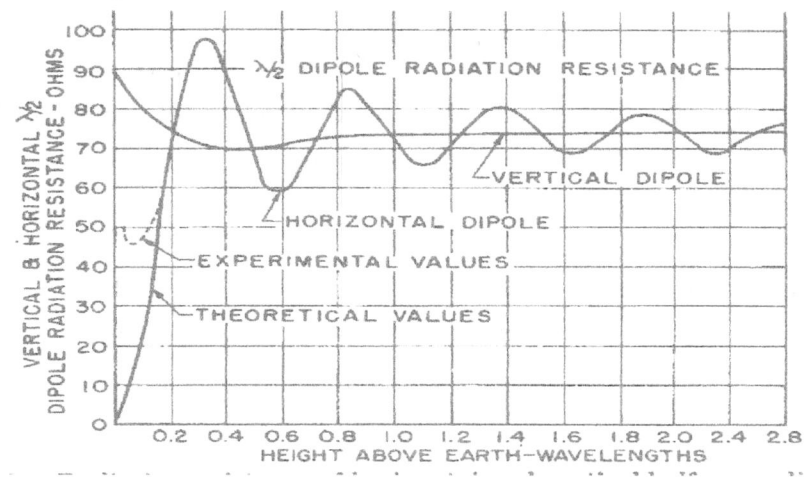

5.9 The Maximum Gain Dipole

If a dipole antenna is extended beyond 1/2-wave, up to a point, its broadside gain will increase. The increase in gain is due to a more narrow beam width. As the length increases beyond the maximum gain length the dipole develops multiple lobes, power is lost from the main broadside lobe to create other lobes.

Maximum broadside gain in a dipole occurs at about 5/4-wave, 1.25 wavelengths. Since the length of a wave in feet = 984/f, multiplying this by 1.25 gives the length in feet of the maximum gain dipole as, 1230/f, or in meters $(300)(1.25) = 375/f$.

This maximum gain dipole is also known as an "Extended Double Zepp" (EDZ) antenna. A double Zepp antenna is a 1-wave length antenna (two half wave co-linear dipoles), hence the "extended" at 1.25 wavelengths.

Unlike a 1/2-wave dipole, the EDZ center reactance will not be zero. And, the radiation resistance will not be 73.2Ω in free space. The feed point impedance of the EDZ, like that of a longer than 90 degree monopole, will have a higher feed point resistance and a reactance.

The EDZ is a 0.625 (5/8) wave parallel transmission line opened out to form a dipole. Being longer than 1/2-wave the EDZ will have a reactance at its feed point. The reactance at the feed point can be calculated using the antenna as transmission line model.

Determining the amount and type of reactance at the feed point of an EDZ becomes simple if we think about a 5/8 wave transmission line. The 5/8 wave line is a 1/2-wave line (4/8) with an added 1/8-wave. The 1/2-wave will repeat the open at the transmission line (antenna) open end. The remaining 1/8 wave will have a capacitive reactance as will any open less than 1/4-wave transmission line.

The reactance of an open 1/8-wave line will be capacitive and equal to the characteristic impedance of the transmission line. The EDZ feed point resistance will be around 150Ω in free space.

As a specific example, an EDZ dipole for 14.1MHz will be, 1230/14.1 = 87.2ft (26.6m), and, like most dipoles, center fed.

For a wire (0.0641 inch, 1.6mm diameter) 20M EDZ, in free space, software calculated, $Z = 147.7 - j823.8Ω$.

A radiation resistance of around 150Ω for free space is a fair approximation for the EDZ. What about the capacitive reactance? The capacitive reactance at the feed point will be $-jX = Za$, the characteristic impedance of the antenna.

Let's calculate the Za for 1/8-wavelength of the 20M EDZ. Since the EDZ is a 5/8-wave transmission line (87.2ft)/(2), 1/8-wave is simply (43.6ft)/(5) = 8.72ft. For wire, Za = 120[Ln(2)(8.72ft)(12in/ft)/(0.0641in)-1] = 851Ω.

Therefore $-j851\Omega$ will be the capacitive reactance at the feed point of the EDZ, found by the antenna as transmission line model.

The estimated free space feed point impedance of the EDZ is, $Z = 150-j\ 851\Omega$ using the transmission line model, which compares well with software's $Z = 147.7-j823.8\Omega$.

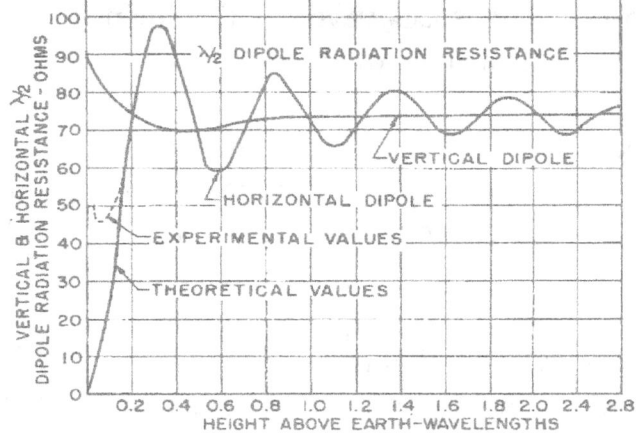

As with any dipole, height above ground will effect the EDZ radiation resistance. For the same 20M wire EDZ at 30ft (9.14m) above real ground, software calculated, $Z = 188.0-j848.2\Omega$.

If we refer to our old friend, the graph of dipole radiation resistance vs height above ground, we can estimate Rr for the EDZ at 30ft above ground too.

A height of 30ft on 20M works out to around 0.43 wavelengths above ground. The graph for a 1/2-wave dipole appears to be about 88Ω, 1.2 times 73.2Ω. We multiply our 150Ω free space Rr approximation for the EDZ by the 1.2 for the changed radiation resistance of the EDZ above ground. And, recalculating the Za for the dipole above ground (using -0.75, not -1), the final, calculator result is, $Z = 180-j881\Omega$, compared to software's $Z = 188.0-j848.2\Omega$.

To bring the radiation resistance closer to 50Ω and cancel the feed point capacitive reactance, a specific length parallel open wire line as an impedance transformer (sometimes called a linear transformer) is often used with an EDZ.

The added specific length of parallel wire line becomes part of the antenna as a system at 14.1MHz, and will have various effects at other frequencies.

Like any dipole fed with a parallel wire feeder the EDZ can be used with a tuner on other frequencies. It will only have maximum broadside gain at a frequency where it is around 1.25 wavelength. At frequencies above the EDZ length the dipole becomes multi-lobed with less broadside gain.

At frequencies below the EDZ length it is much like any regular 1/2-wave dipole having some broadside gain with broader lobes and wider beam width.

When selecting a length for a parallel line fed general use dipole some choose to use an EDZ at a certain frequency for its maximum gain at that frequency. As an example, an EDZ for 10M, around 44ft (13.4m) long, would give max gain for 10M and a regular dipole pattern at lower frequencies. Or, if space will allow, put up a 20M EDZ, of 87ft (26.6m) length for maximum broadside gain in that band and have a dipole pattern at lower frequencies and a multi-lobe pattern at higher frequencies.

This maximum gain is also why 5/8-wave verticals are used, which amounts to half an EDZ dipole with a perfect ground. The 5/8-wave vertical monopole has been very popular at VHF. The increase in gain over a 1/4-wave monople means the angle of radiation is lowered for mobile to mobile and mobile to repeater communications.

As we now know, an EDZ will have a capacitive reactance at its feed point. The 5/8-wave vertical monopole, being half an EDZ dipole, will have a capacitive reactance at the feed point, and that is why it will have a "loading coil" at its base.

What will be the feed point impedance of the 5/8-wave monopole? Based upon the examinations in the section "The Dipole, Two Monopoles in Series – The Image Becomes Real", the feed point impedance over perfect ground of a 5/8-wave monopole should be half that of the dipole in free space.

For the wire EDZ dipole in free space at 14.1MHz, software calculated, $Z = 147.7 - j823.8\Omega$, half would be $Z = 73.85 - j411.9\Omega$. For a wire 5/8-wave monopole over perfect ground, software calculated, $Z = 72.96 - j408.7\Omega$.

By placing an inductor in series with the 5/8-wave monopole at the feed point the capacitive reactance of the antenna can be canceled. Adding a 4.6µH inductor, acting as a base loading coil, to the antenna model of the 5/8-wave wire 20M monopole, software calculated, $Z = 72.71 - j1.258$, 1.45 SWR.

The ultimate in a single element, vertically polarized, low take-off angle, gain antenna, is a 5/8-wave monopole over a perfectly conducting, infinite, ground plane.

There are some sources on the topic of the Extended Double Zepp that use a total length of 1.28, rather than 1.25 wavelengths. Here is an example from U.S. Army manual TM 11-666 "Antennas and Radio Propagation" from 1953 using 1.28 wave length. This is why I wrote "a length around 1.25 wavelengths" in my opening remarks.

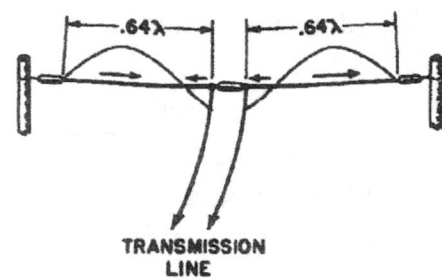

EXTENDED DOUBLE-ZEPP ARRAY

In an April 1935 IRE paper, "General Considerations of Tower Antennas for Broadcast Use" Gihring & Brown found that for a monopole over a ground radial system, the maximum ground level radiated field strength occurred at a tower height of 0.64 wave length. Using their result for a monopole, a dipole would need to be twice as long, 1.28 wave length, matching the 1953 Army training manual value.

A software gain sweep of an EDZ dipole indicates a maximum gain of +5.18dBi at 1.269 wavelength in free space, which would be 0.6345 for a monopole, pretty much 0.64. A gain of +5.18dBi is +3.3dB more than the basic 1/2-wave dipole (+3dB is a doubling of power). This power gain is due to the EDZ narrow beam width, only true for free space, approximately true for a height of 1/2-wave above ground.

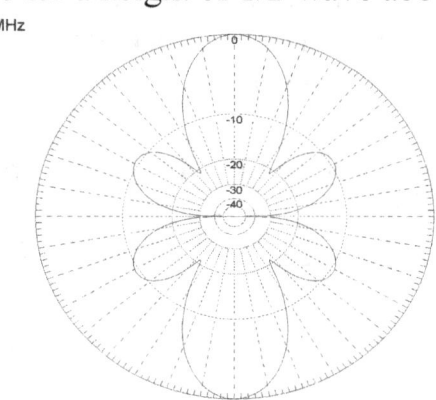

EDZ 20M 14.2MHz

The G5RV Dipole

Another example of a longer than 1/2-wave, multi-lobed, center fed dipole is the 1.5-wave (3/2-wave) dipole. This length antenna was made popular by G5RV as a center fed directional dipole for 14.2MHz. At 1.5-wave the center feed point radiation resistance is around 106Ω in free space.

To find the length of a 3/2-wave antenna in feet, (3)(984/2) = 1476/f, which at 14.2MHz is a length of 1476/14.2 = 104ft (31.7m) (typically 102ft for the G5RV). By going beyond the length of the EDZ the 1.5-wave dipole has four main lobes, six lobes in total.

119

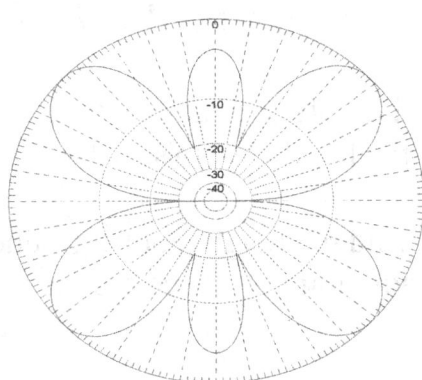

The G5RV is another example of a purpose length, longer than 1/2-wave dipole. When fed with a parallel feeder to a tuner the G5RV can be used on other frequencies, just as the EDZ can.

As the free space pattern of the G5RV shows, it has 4 major lobes and the two broadside lobes have gotten smaller compared to the EDZ. The pattern would be approximately the same if the G5RV were at 1/2-wave above ground.

The G5RV can be positioned so the 4 major lobes are in the direction of four compass points of interest. The gain of those major lobes are about +1.32dB over a 1/2-wave dipole. The two minor lobes have less gain than a 1/2-wave dipole. As with any dipole, height above ground will effect the feed point resistance of the G5RV.

I find it interesting that the 1936 W3EDP off-center fed antenna suggested 17ft for one length of the counterpoise wire. With 84ft of "antenna" wire and 17ft counterpoise, the total length becomes 101ft, very close to 102ft of the G5RV 3/2-wave dipole.

Chapter 6 Vertical Dipoles and the Vertically Polarized Loop

Having examined the horizontal dipole and its variations in the last section, now the dipole is rotated to become vertical to the earth's surface.

When the dipole is vertical we find the height above ground effect on antenna characteristic impedance and radiation resistance are much less than for the horizontal dipole.

As with the horizontal dipole, the vertical dipole can be inductively loaded, or it can be capacitively end loaded.

In the monopole sectionss using a counterpoise and end loading the "I" vertical was developed. By capacitively end loading the monopole on both ends an off-center fed vertical dipole resulted. The capacitively end loaded off-center fed vertical dipole will make another appearance in this section.

Two examples of end capacitive loading, including some measured values from constructed antennas are presented for comparison. In one case, both ends of the vertical dipole will be capacitively loaded. In other examples the vertical dipole is capacitively loaded on one end only.

For the final example, a full wavelength vertical loop is presented as two capacitively end loaded vertical dipoles with the end loading wires connected to form a loop. By treating the two end loaded vertical dipoles as interconnected to form a loop, the feed point radiation resistance of the loop can be selected by the height of the vertical conductors and then modeled as a 1/2-wave transmission line.

6.1 The Vertical Dipole

Advantages of the vertical dipole over horizontal are the radiation resistance is much less variable for height above ground and is greater than either a 1/4-wave monopole or 1/2-wave horizontal dipole.

It would be fair to object and point out that the height of a vertical dipole could exceed the height at which a horizontal dipole might be hoisted at low frequencies. As you will see, the higher and more stable radiation resistance of the vertical dipole makes possible physical shortening that could not be done with a monopole or low horizontal dipole, yet achieve a low SWR with a 50Ω feeder.

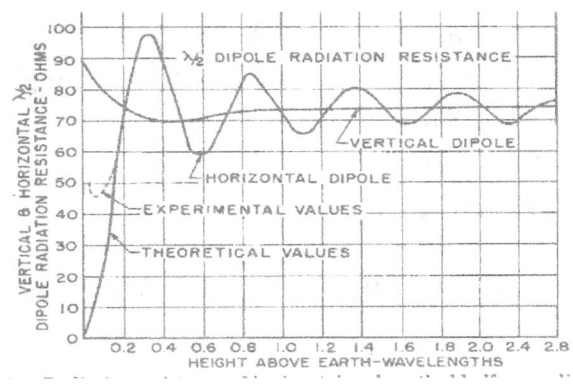

As the graph shows, when the bottom of a vertical dipole is just above ground level, its radiation resistance approaches around 90Ω. Long before computers the radiation resistance of a (thin) vertical dipole above ground was calculated by one source to be 98.5Ω. For a resonant conduit vertical dipole, about 1ft (0.3048m) above ground, software calculated Rr = 97Ω.

We can lose half the nearly100Ω radiation resistance in shortening and still have a 50Ω, 1:1 SWR. A vertical dipole can be 75% of its resonant 1/2-wave length and provide a radiation resistance around 50Ω. Being short it will have a capacitive reactance at its feed point, but we have solutions for that. Being able to shorten the vertical dipole to 3/4-length and still achieve a low SWR, with loading, is one of the reasons I have gravitated toward vertical dipoles over the years, that and vertical polarization with no need for a ground system.

As an example, a 20ft (6.10m) conduit vertical dipole is very close to 3/4-length at 18.1MHz and will provide a radiation resistance around 50Ω at that frequency.

The Za = 120[Ln(2)(10ft)(12in/ft)/(0.922in) -1] = 547Ω. Notice I am back to using -1 and not -0.75 in the Za formula for the vertical dipole. For a vertical dipole there is no need to change the free space dipole Za formula.

The length in degrees of the 20ft conduit dipole at 21.1, 18.1 and 14.1MHz.
     (21.1)/(23.4)(90 degrees) = 81.15 degrees.
     (18.1)/(23.4)(90 degrees) = 69.62 degrees.
     (14.1)/(23.4)(90 degrees = 54.23 degrees.

Using transmission line model the calculated capacitive reactances at the center feed point of the 20ft conduit vertical dipole will be,
     21.1MHz   (547Ω)/(tan 81.15) = -j85.2Ω.
     18.1MHz   (547Ω)/(tan 69.62) = -j203Ω
     14.1MHz   (547Ω)/(tan 54.23) = -j394.

Estimation of the radiation resistance and SWR (with the capacitive rectance canceled) using $Rr = Ro[(1-cos\Theta)^2/sin^2\Theta]$ where $Ro = 100\Omega$ for a vertical dipole is:

21.1MHz   73Ω   1.46 SWR.
18.1MHz   48Ω   1.04 SWR.
14.1MHz   26Ω   1.92 SWR.

Here are the software values for the 20ft center fed conduit vertical dipole, 1ft above ground for comparison,

21.1MHz   Z = 74.91-j80.56Ω.
18.1MHz   Z = 52.48-j198.8Ω.
14.1MHz   Z = 30.91-j392.3Ω.

The radiation resistance SWR for each, using the software Rr values will be,

21.1MHz   75/50 = 1.5 SWR.
18.1MHz   53/50 = 1.1 SWR.
14.1MHz   50/31 = 1.6 SWR.

Canceling the capacitive reactance at the feed point can be accomplished with inductive loading, either with coils or shorted inductive stubs. Let's use wire with 2 inch (50.8mm) spacing to form a parallel transmission line stub as an example. The characteristic impedance of the line will be, $Zo = 120Ln(2)(2in)/(0.0461in) = 496\Omega$.

At 21.1MHz we need +j85.2Ω. The transmission electrical length needed will be $\Theta = tan^{-1}(85.2\Omega)/(496\Omega) = 9.747$ degrees. The length of a wave in free space is, (984)/(21.1MHz) = 46.64ft. The physical length of the shorted stub will be (9.747 degrees)/(360 degrees)(46.64ft) = 1.26ft (0.384m).

At 18.1MHz we need +j203Ω. The electrical length needed will be $\Theta = tan^{-1}(203\Omega)/(496\Omega) = 22.26$ degrees. The length of a wave in free space is (984)/(18.1MHz) = 54.36ft. The physical length of the shorted stub will be (22.26 degrees)/(360 degrees)(54.36ft) = 3.36ft (1.02m).

At 14.1MHz we need +j394Ω. The electrical length needed will be $\Theta = tan^{-1}(394\Omega)/(496\Omega) = 38.46$ degrees. The length of a wave in free space is (984)/(14.1MHz) = 69.79ft. The physical length of the shorted stub will be (38.46 degrees)/(360 degrees)(69.79ft) = 7.46ft (2.27m).

There you have it, a 20 ft. vertical dipole for three popular HF bands, with low (less than 2:1) SWR.

A vertical dipole has a lower radiation angle compared to a monopole over perfect ground. This lower angle also comes with greater gain. No ground radial system is required with a vertical dipole.

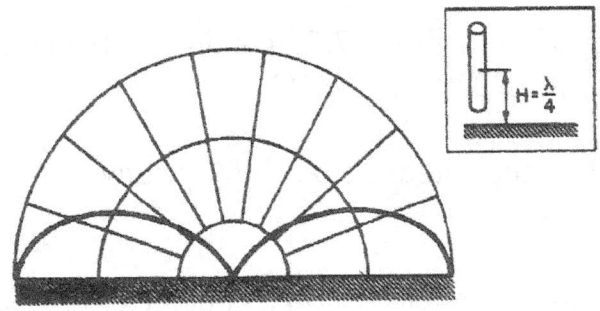

Vertical dipole radiation pattern with its bottom end very near ground.
From U.S. Army TM 11-666 "Antennas and Radio Propagation" 1953

A later example in this section will show how to cancel the capacitive reactance by placing the inductive loading at other than the feed point.

6.2 An Octave Vertical Dipole

In electronics, as in music, an octave is a doubling of frequency.

Thanks to the higher and more constant radiation resistance of the vertical dipole it is possible for a vertical dipole to cover an octave with a low SWR.

To accomplish an octave coverage from a vertical dipole the length needs to be around 70 degrees at the center of the octave range. The 70 degree length will be 0.8 times (70 degrees/90 degrees) the calculated physical length at the middle of the octave range. This multiplier of 0.8 at the center of the octave range is related to the radiation resistance following sine and cosine rules. Remember the "sweet spot" for the monopole? A similar concept is at work here with the vertical dipole.

To find the physical length of a vertical dipole to cover from 20M to 10M, an octave frequency range, start by calculating a 1/2-wave dipole at the center frequency of the selected range. For this example I will use 14.1 to 28.2MHz. The center of this range is 21.15MHz.

The physical length of a dipole at this center frequency will be, (468/21.15) = 22.13ft. The final length of the vertical dipole will be (22.13ft)(0.8) = 17.7ft (5.40m).

The natural resonant frequency of this dipole will be, (468/17.7ft) = 26.4MHz.

I will now model the antenna as a transmission line to find the reactance needed at the center feed point for conjugate matching, 20M through 10M.

First the characteristic impedance of the 17.7ft conduit vertical dipole
$Za = 120[\ln(17.7ft)(12ft/in)/(0.922) -1] = 533\Omega$.

Using the electrical length of the antenna at each frequency and the basic transmission line formula, $-jX = (Za)/(\tan\Theta)$:

14.1MHz $\quad\Theta = (14.1/26.4)(90 \text{ degrees}) = 48.07 \text{ degrees}, -j479\Omega$.
18.1MHz $\quad\Theta = (18.1/26.4)(90 \text{ degrees}) = 61.71 \text{ degrees}, -j287\Omega$.
21.1MHz $\quad\Theta = (21.1/26.4)(90 \text{ degrees}) = 71.93 \text{ degrees}, -j174\Omega$.
24.9MHz $\quad\Theta = (24.9/26.4)(90 \text{ degrees}) = 84.89 \text{ degrees}, -j47.7\Omega$.
28.2MHz $\quad\Theta = (28.2/26.4)(90 \text{ degrees}) = 96.14 \text{ degrees}, +j57.3\Omega$.

Using the relationship $Rr = Ro[(1-\cos\Theta)^2/(\sin\Theta)^2]$, and $Ro = 100\Omega$, the electrical lengths and reactance calculated above, the estimated and software feed point impedances are:

14.1MHz $\quad Z = 20-j479\Omega \quad$ Software $\quad Z = 23.92-j480.1\Omega$.
18.1MHz $\quad Z = 36-j287\Omega \quad$ Software $\quad Z = 39.9-j284.3\Omega$.
21.1MHz $\quad Z = 53-j174\Omega \quad$ Software $\quad Z = 55.69-j170.7\Omega$.
24.9MHz $\quad Z = 84-j47.7\Omega \quad$ Software $\quad Z = 82.71-j45.02\Omega. \quad$ 2.32 SWR, as-is.
28.2MHz $\quad Z = 124+j57.3\Omega \quad$ Software $\quad Z = 116.6+j58.99\Omega. \quad$ 3.03 SWR, as-is.

The SWR on 10M and 12M are low enough to not require conjugate matching. I will leave them as they are.

For the other frequencies inductive reactance will be used. I will use my transmission line calculated reactance to find the inductance values for feed point loading:
14.1MHz $\quad L = 479\Omega/2\pi(14.1) = 5.4\mu H$.
18.1MHz $\quad L = 287\Omega/2\pi(18.1) = 2.5\mu H$.
21.1MHz $\quad L = 174\Omega/2\pi(21.1) = 1.3\mu H$.

Using my calculated inductance values, software gives the following:
14.1MHz $\quad Z = 23.92-j1.737\Omega \quad$ 2.09 SWR.
18.1MHz $\quad Z = 39.9-j0.0093\Omega \quad$ 1.25 SWR.
21.1MHz $\quad Z = 55.69+j1.639\Omega \quad$ 1.12 SWR.

One coil with three tap points could be a handy solution for these three bands. Bypass the inductance for operation on 10M and 12M.

Or, as in the previous chapter, shorted stub(s) could be used rather than coils to provide the needed inductive reactance.

Another possible octave vertical dipole is 40 to 20M, which would cover 40, 30 and 20 meter bands. The same process applies, a dipole with a final physical length of 0.8 times the calculated physical length of the middle frequency.

The middle frequency will be, (7+14.35)/2 = 10.7MHz. The physical vertical dipole length will be, (468/10.7MHz)(0.8) = 35ft (10.67m). The Za of a 35ft conduit dipole is, 615Ω. The self resonant frequency is, (468)/(35ft) = 13.4MHz.

Using the methods shown in the previous example and a calculator I got the following:

14.1MHz	94.70 degrees	Z = 118+j50.6Ω.
10.12MHz	67.97 degrees	Z = 45.4-j249Ω.
7.1MHz	47.69 degrees	Z = 19.5-j560Ω.

Software calculated:

14.1MHz	Z = 114.9+j57.05Ω	2.96 SWR "as is."
10.12MHz	Z = 50.96-j249.0Ω.	
7.1MHz	Z = 24.3-j572.7Ω.	

The software, using my calculated inductance for 10.12MHz and 7.1MHz, gives the following:

| 10.12MHz | L = 3.9µH | Z = 50.96-j1.03Ω | 1.03 SWR. |
| 7,1MHz | L = 12.8µH | Z = 24.3-j1.661Ω | 2.06 SWR. |

All three bands, two inductors, or one inductor with a tap, an SWR of less than 3:1 from one 35ft center fed conduit vertical dipole 1ft above ground.

A monopole for 7.1MHz would have a height of 33ft (10m), require an extensive ground system and have a high impedance on 20M. A 35ft (10.7m) vertical dipole, just 2ft more and 1ft above ground, will give a low SWR on 7.1MHz, 10.12MHz and 14.1MHz, no ground system required. Or if fed with parallel line to a tuner, all HF bands.

Are you becoming a fan of vertical dipoles yet?

In a later example I will show why I became a fan of the vertical dipole for a stealth antenna.

6.3 A Capacitive End Loaded Vertical Dipole

The higher radiation resistance of the vertical dipole makes possible physical shortening using capacitive end loading and still provide multi-band operation.

In this chapter I will end load a 15ft (4.57m) conduit vertical dipole with two 13 inch (0.33m) pizza pans. As before, I will make antenna as transmission line calculations for comparison to a software model. I also have measured values from a constructed antenna for comparison.

First the 15ft vertical dipole without end-loading.

For a 15ft, 0.922in (21mm) diameter conduit vertical dipole the $Za = 120[Ln(2)(7.5ft)$ $(12in/ft)/(0.922in) -1] = 513\Omega$. The calculated natural resonant frequency will be $(468/15) = 31.2MHz$.

The length in degrees of the 15ft conduit dipole at 28.2, 24.9, 21.1, 18.1 and 14.1MHz:
$(28.2/31.2)(90 \text{ degrees}) = 81.35$ degrees.
$(24.9/31.2)(90 \text{ degrees}) = 71.83$ degrees.
$(21.1/31.2)(90 \text{ degrees}) = 60.87$ degrees.
$(18.1/31.2)(90 \text{ degrees}) = 52.21$ degrees.
$(14.1/31.2)(90 \text{ degrees}) = 40.67$ degrees.

Since all the electrical lengths are less than 90 degrees, all the feed point reactance values will be capacitive. Next I calculate the feed point capacitive reactance and the estimated radiation resistances for each frequency.

First the reactance at each frequency using the standard formula for any less than 1/4-wave open transmission line, $-jX = Za/tan\Theta$:

28.2MHz	$-j78.1\Omega$
24.9MHz	$-j168\Omega$
21.1MHz	$-j286\Omega$
18.1MHz	$-j398\Omega$
14.1MHz	$-j597\Omega$

Now the estimated radiation resistance at each frequency using $Rr = Ro[(1-cos\Theta)^2/sin^2\Theta]$. Previously I used 100Ω for Ro, I will amend that to the 97Ω found by software for conduit:

Calculated	Software	
28.2MHz	Z = 72-j78.1Ω	Z = 69.95-j79.89Ω
24.9MHz	Z = 51-j168Ω	Z = 52.58-j169.9Ω
21.1MHz	Z = 34-j286Ω	Z = 37.08-j287.9Ω
18.1MHz	Z = 23-j398Ω	Z = 27.2-j401.2Ω
14.1MHz	Z = 13-j597Ω	Z = 16.64-j606.1Ω
15.6MHz		Z = 20.36-j520.2Ω

The frequency 15.6MHz is half the resonant frequency of 31.2MHz. At half resonant frequency the antenna is a 1/8-wave open transmission line. The reactance of any 1/8-wave transmission line (whether open or shorted) is equal to the characteristic impedance (Zo) of that line, which for an antenna will be Za, the characteristic impedance of the antenna as a transmission line. My calculated Za of 513Ω compares well with the software value of 520.2Ω.

Here are the measured values from the physical antenna, with the bottom end of the vertical dipole approximately 18 inches (0.457m) above ground (RigExpert AA-55 Zoom):

28.2MHz	Z = 76.2-j76.3Ω
24.9MHz	Z = 59.4-j167.4Ω
21.1MHz	Z = 32.0-j259.6Ω
18.1MHz	Z = 31.4-j368.4Ω
14.1MHz	Z = 5.23-j597.4Ω

The measured resonant frequency was 30.8MHz, making the 1/8-wave test frequency 15.4MHz, resulting in a measured -jX = Za = 508Ω, which is different than the software value, but very close to my calculated value of 513Ω.

We now have comparisons between the calculated by transmission line model, the software model, and measuring a physical 15ft conduit vertical dipole antenna.

Capacitive End-loading

Next I added 13 inch pizza pans to each end of the 15ft conduit vertical dipole. As calculated in previous section, with monopoles, the capacitive end-loading will increase the electrical length of the antenna, lowering the resonant frequency.

Pizza pans added to both ends of the 3/4-inch diameter, 15ft conduit vertical dipole

The capacitance provided by the 13 inch solid disk is, (13inches)(0.8992) = 11.7pF. The reactance of 11.7pF at the 15ft vertical dipole calculated resonate frequency of 31.2MHz will be, $-jX = 1/(2)(\pi)(31.2\text{x}10^6)(11.7\text{x}10^{-12}) = -j436\Omega$.

Now to find the equivalent electrical length of $-j436\Omega$ due to the disk end loading when added to the antenna as a transmission line vertical dipole with $Za = 513\Omega$.

The added electrical length in degrees will be, $\Theta = \tan^{-1}(Za)/(-jX) = \tan^{-1}(513\Omega)/(436\Omega)$ = 49.64 degrees. To find the physical length of this calculated electrical length we need the free space length of a wave at 31.2MHz which will be (984)/(31.2) = 31.54 feet. The equivalent physical length of the pizza pans is, (49.64 degrees/360 degrees)(31.54ft) = 4.35ft (1.33m). This length, added to the dipole 15ft physical length, gives a total equivalent physical length of 19.35ft (5.60m). The new calculated resonate frequency is, (468)/(19.35ft) = 24.2MHz.

Using the antenna as transmission model the electrical length of a 19.35ft conduit vertical dipole:

	Calculated	Software
28.2MHz	Z = 167+j145Ω	Z = 147.7+j145.2Ω
24.9MHz	Z = 106+j24.7Ω	Z = 100.3+j23.45Ω
21.1MHz	Z = 65-j111Ω	Z = 65.93-j114.7Ω
18.1MHz	Z = 43-j227Ω	Z = 46.91-j233.9Ω
14.1MHz	Z = 24-j418Ω	Z = 28.09-j432.0Ω

The measured resonant frequency was 23.9MHz, 1.2% lower than the calculated value, for a physical length of (468/23.9MHz) = 19.58ft. Not bad for pencil, paper and calculator.

Here are the measured values from the antenna analyzer:

28.2MHz	$Z = 173.9+j158.5\Omega$
24.9MHz	$Z = 113.3+j94.6\Omega$
21.1MHz	$Z = 62.6-j103.6\Omega$
18.1MHz	$Z = 42.5-j219.2\Omega$
14.1MHz	$Z = 19.5-j418.7\Omega$

As with the 20ft vertical dipole of the last chapter, this 15ft vertical dipole, end loaded with two 13in pizza pans, is an octave antenna. The measured SWR at 24.9MHz was 2.3:1 which would be fine as is. The other frequencies can be conjugate matched at the feed point using a capacitor for 28.2MHz, and inductors (or one inductor with taps), or shorted stubs, for 21.1, 18.1, and 14.1MHz.

Using the measured radiation resistance values, the SWR values after conjugate matching would be:

28.2MHz	173.9/50 = 3.5:1 SWR
24.9MHz	as is 2.3:1 SWR
21.1MHz	62.6/50 = 1.3:1 SWR
18.1MHz	50/42.5 = 1.2:1 SWR
14.1MHz	50/19.5 = 2.6:1 SWR

All measured values were made with a bead type (W2DU) broadband choke balun at the feed point.

The capacitive end loading of a vertical dipole using disks in this example is a symmetrical case. An asymmetric example will be examined next.

6.4 Vertical Dipole Asymmetrical Capacitive End Loading

While I had the 15ft center fed conduit vertical dipole up for impedance measurements, before the addition of the pizza pans, I decided to add two equal length wires at the bottom end of the vertical dipole, near ground, as capacitive end loading.

The natural resonant frequency of the vertical dipole should be, (468/15) = 31.5MHz.
The wave length at that frequency is (984/31.2MHz) = 31.54ft (4.57m).
The conduit vertical dipole characteristic impedance, Za = 513Ω, calculated in the previous section.

Two equal length elevated wires as a capacitive ground

I used a wire scrap 10ft (3.05m) long, 1ft (0.305m) above ground, and connected the wire center to the bottom end of the conduit dipole. This wire added two, 5ft (1.52m), single wire transmission line capacitive loads to the bottom end of the conduit vertical dipole. The feed point remained at the center of the 15ft conduit vertical dipole.

What will be the new, lower, resonant frequency with the added wire? I will measure it, but I wanted to calculate the new resonant frequency using the antenna as transmission line method.

The Zo of the wire 1 ft above ground has previously been calculated to be 397Ω.
Using Zo = 397Ω for the horizontal wires, I calculated the electrical length, in degrees, for 5ft (1.52m) of wire at the vertical dipole resonant frequency,
(5ft/31.54ft)(360degrees) = 57.07 degrees. The capacitive reactance of this open transmission line is -jX = (397Ω)/(tan 57.07 degrees) = -j257Ω.

At the bottom end of the vertical dipole the two 5ft wires make a parallel connection, the total capacitive reactance will be half -j257Ω, -j129Ω. This capacitive reactance will add length in degrees to the 15ft conduit antenna, $\Theta = \tan^{-1}(513\Omega)/(129\Omega) = 75.89$ degrees. The added length in feet will be (75.89 degrees/360 degrees)(31.54ft) = 6.65ft (2.04m).

The new, lower resonant frequency will be (468)/(15ft + 6.65ft) = 21.6MHz. The measured minimum SWR was 2.2:1 at 21.66MHz, with a feed point impedance (center of the 15ft conduit vertical dipole) of Z = 96.8-j27.9Ω.

Adding capacitance to one end of a vertical dipole, not only lowers the resonant frequency when center fed, it also creates an off-center fed vertical dipole.

Off-center feed causes the resistance at the dipole current maximum to be multiplied. This was seen with the Windom dipole.

Here are software calculated feed point impedance values:

21.1MHz	Z = 79.72-j57.35Ω	2.67 SWR as is
18.1MHz	Z = 52.17-j191.2Ω	
14.1MHz	Z = 28.61-j400.4Ω	
10.12MHz	Z = 14.01-j709.3Ω	
7.1MHz	Z = 6.863-j1132Ω	

At this point you might expect me to say that with feed point (the center of the conduit vertical dipole) inductive loading this antenna would have a low SWR on 15, 17 and 20M, and it would. The thing is, there is a better place to put the inductive loads.

That better location is at the connection between the bottom end of the conduit vertical dipole and the capacitive ground wires, rather than at the feed point. But how much inductance will be required for off-set loading at that location?

With an antenna analyzer this is a very easy question to answer. Simply change the feed point from the center of the conduit to the bottom end of the conduit where it is connected to the wire capacitive ground, and place a short at the conduit center feed point. (Remember, when sources are removed a short is put in their place?)

Here are the software calculated impedance values at that location:

21.1MHz	Z = 92.03-j81.84Ω.
18.1MHz	Z = 52.17-j191.2Ω.
14,1MHz	Z = 28.61-j400.4Ω
10.12MHz	Z = 14.01-j709.3Ω.
7.1MHz	Z = 4.189-j1107Ω.

The reactance values are those required for off-set loading of the vertical dipole. The antenna will still be fed at the center of the 15ft conduit vertical dipole, but the inductive loading will not be at the feed point. The loading will be applied where the bottom of the conduit vertical dipole meets the counterpoise wires.

Using the capacitive rectance values at each frequency provided by software I calculated the inductance, $L = Xc/(2\pi)(Fo)$, required to cancel the calculated capacitive reactance at the loading location.

Here are the final software results for the center fed conduit vertical dipole with an off-set inductive load connected from the bottom of the conduit dipole to the two capacitive ground wires:

21.1MHz	$Z = 94.25+j8.875\Omega$	1.91:1 SWR	$0.617\mu H$
18.1MHz	$Z = 80.25+j2.395\Omega$	1.61:1 SWR	$1.95\mu H$
14.1MHz	$Z = 57.99-j0.1216\Omega$	1.16:1 SWR	$4.8\mu H$
10.12MHz	$Z = 34.47+j6.036\Omega$	1.49:1 SWR	$11.28\mu H$
7.1MHz	$Z = 18.1-j6.149\Omega$	2.81:1 SWR	$24.8\mu H$

Four bands with SWR less than 2:1 and the fifth (40M) less than 3:1 (more than an octave!). The bandwidth on 40M will be very narrow due to the large -jX which will result in a high Q antenna, the fate of all physically short antennas. The same will be true for 30M but it is a narrow band anyway.

The low SWR values of this vertical dipole are possible for two reasons. The first is the feed point may be fixed, but as the frequency lowers, that fixed location from the top end of the conduit increases the off-set, thus increasing the multiplier effect of the dipole radiation resistance. The conduit becomes an off-center fed inverted "T" vertical dipole, and more so as the frequency lowers.

The second reason is due to the inductive loading not being at the feed point. As we learned with the monopole and dipole, moving the inductive load from the feed point out toward the end of the radiator caused the radiation resistance to increase.

This asymmetrical capacitive end loaded vertical dipole has the advantage of off-set feed, and off-set inductive loading. An adjustable inductor makes this antenna multi-frequency and low SWR. The inductor is placed in series with the conduit and the wire counterpoise at the dipole bottom end.

I have constructed, for portable use, a version of this antenna using a wire vertical dipole and an adjustable inductor between the bottom of the vertical dipole and the wire capacitive counterpoise. Eventually I eliminated the bottom counterpoise wires replacing them with a capacitive disk (pizza pan), making the entire antenna more compact.

6.5 The Flagpole Loophole Vertical Dipole

A move, after retirement, landed me in a community with many property use restrictions including antennas. This simply would not do for an HF operator like myself. I started with low horizontal "stealth" loops, the topic of the next section.

A careful reading of the rules revealed that a flagpole was allowed. The rules specified type (white or metal) and even specified the flags that could be displayed. The flagpole could have a maximum height of 20ft (6.01m). A search on the internet found a 20ft metal flagpole kit that came in 4ft sections with all the attachments, the flag, even a ball on top.

A radial system for a monople was really not possible at my location (plus I don't do radials, too much wire, too much work, I'm retired). I decided on a vertical dipole. Feeding a vertical dipole at the center of a flagpole was not really possible. I decided to make an inverted T vertical dipole and feed the flagpole at the base using a remote tuner.

The entire vertical dipole consists of the 20ft, 2in (50.8mm) diameter, metal flagpole, and 10 feet of 1/2in (12.7mm) copper pipe placed horizontally at the base to form a capacitive end load. The center conductor of the coax runs to the flagpole, the coax shield to the center of the 10ft (3.05m) copper pipe at the bottom end of the 20ft flagpole.

This arrangement results in an inverted T, off-center fed vertical dipole. Since the horizontal copper pipe is center connected the currents should be equal, yet opposite, and cancel, resulting in little to no horizontal radiation.

Though the antenna is to be used with an automatic tuner at the feed point, I wondered what the self resonant frequency would be. Out came the pencil, paper and calculator. Later, the computer software for comparison.

Treating the flagpole as a dipole, the self resonant frequency of the flagpole alone is, (468)/(20ft) = 23.4MHz. The wavelength will be (984)/(23.4MHz) = 42.05ft (12.8m). The characteristic impedance of the flagpole dipole as a transmission line is Za = 120[Ln(2)(10ft)(12in/ft)/(2in) − 1] = 455Ω.

The remainder of the dipole is the center connected 10ft horizontal copper pipe, about an average 8 inches (0.20m) above ground, its Zo = 60[Ln(4)(8in)/(0.63in)] = 236Ω. The electrical length of 5ft (1.52m) (each half of the 10ft copper pipe) is (5ft/42.05ft)(360 degrees) = 42.81 degrees.

The capacitive reactance of the 5ft of copper pipe is $-jX = (Zo)/(\tan\Theta) = (236\Omega)/(\tan 42.81) = -j255\Omega$. Since the two 5ft lengths are connected in parallel at the base (bottom end) of the flagpole, the total reactance will be $(-j255)/(2) = -j128\Omega$. This capcacitive reactance is added to the bottom end of the 20ft flagpole vertical dipole. Now to find the new electrical length and from that the new physical length of the flagpole dipole.

The electrical length of the copper pipe will be $\Theta = \tan^{-1}(Za)/(-jX) = (455\Omega/128\Omega) = 74.29$ degrees of added length. The physical length will be (74.28 degrees/360 degrees) (42.05ft) = 8.68ft added to the 20ft flagpole for a total equivalent physical length of (20ft + 8.68ft) = 28.7ft (8.76m). The new resonant frequency of the vertical dipole (468/28.7ft) = 16.3MHz.

This vertical dipole is off-center fed. This places the feed point at a higher impedance point than if it had been center fed as discussed in "The Off-Center Fed Horizontal Dipole" section and the asymmetrical vertical dipole in the previous section.

The multiplier is related to the distance from the center of the dipole (current maximum) to the new location by $Rr/\cos^2\Theta$. Or, from the dipole end, by $Rr/\sin^2\Theta$.

Since the previously calculated 8.7ft of the 10ft copper pipe is the equivalent added length, $\Theta = (8.7\text{ft}/14.35\text{ft})(90\text{ degrees}) = 54.56$ degrees. The multiplier will be $Ro/\sin^2\Theta$. Usually $Ro = 97\Omega$ for a vertical dipole at the center, but in this example the dipole is bent. As seen in "The Bent Ends Horizontal Dipole" section the radiation resistance is reduced some due to bending the ends.

Rather than do calculations to discover how much the 97Ω Ro is reduced due to bending I used Ro = 90Ω as my best educated guess. The off-set resistance should be about, $(90\Omega)/(\sin^2 54.56 \text{ degrees}) = 136\Omega$ at resonance. The software model calculated the resonant frequency to be 16.6MHz (compared to 16.3MHz calculated) with a feed point impedance of $Z = 133.5 + j0.8591\Omega$, 2.67:1 SWR.

Or is this really a monopole with a capacitive ground?

In previous sections it was shown that placing a capacitor in series at the feed point electrically shortened a monopole or dipole and raised the resonant frequency. That fact can be applied to this 20ft flagpole by treating it as a monopole rather than as a vertical dipole.

The flagpole as a 20ft (9.096m) monopole over perfect ground has a resonant frequency of about 234/20ft = 11.7MHz. But the flagpole is not using a perfect ground, it has a capacitive ground created by the 10ft copper pipe.

The -j128Ω capacitive reactance of the copper pipe to ground will be in series at the feed point, electrically shortening the monopole, and raising the resonant frequency from 11.7MHz.

The procedure in earlier sections was to select a monopole length greater than 90 degrees so the antenna had an inductive reactance at the feed point and the capacitance of the counterpoise was selected to cancel the monopole inductive reactance.

In the case of the flagpole monopole, the length is fixed at 20ft. Reverse logic would make one conclude that there must be some frequency, higher than 11.7MHz, at which the 20ft flagpole will have the necessary inductive reactance to cancel the capacitive reactance of the copper pipe counterpoise at a new monopole resonant frequency. Following that line of reasoning, at resonance, the radiation resistance should be greater than the 36.6Ω of a 90 degree monopole too. The -j128Ω total capacitive reactance of the horizontal copper pipe counterpoise will *subtract* electrical length from the 20ft monopole because it is added at the feed point.

The characteristic impedance of the 20ft flagpole as a monopole will be
Zo = 60[Ln(2)(20ft)(12 in/ft)/2in -1] = 269Ω. Using the antenna as transmission line model, the new electrical length, Θ = tan^{-1} (Za)/(-jX) = (269Ω)/(128Ω) = 64.55 degrees. The monopole is no longer 90 degrees, and resonant at 11.7MHz. It will be resonant at a higher frequency due to its shortened electrical length by the counterpoise capacitive reactance.

The new physical length will be (64.55 degrees/90 degrees)(20ft) = 14.35ft. The new resonant frequency will be (234/14.35ft) = 16.3MHz when treated as a monopole with a capacitive ground at the feed point.

Because the new electrical length at resonance is 64.55 degrees, the "missing" 25.45 degrees is the portion of the 20ft monopole that contributed +j128Ω inductive reactance to cancel the -j128Ω feed point capacitive reactance of the copper pipe counterpoise.

The estimated radiation resistance for this monopole at 16.3MHz, from the relationship Rr = Ro [(1-cosΘ)^2/sin$^2\Theta$], where Ro for a monopole over perfect ground (ideal) is 36.6Ω and Θ = (16.3MHz/11.7MHz)(90 degrees) = 125.4 degrees for Rr = 137Ω, compared to my off-center fed dipole 136Ω and software's 133.5Ω.

The copper pipe counterpoise added 45.19pF to the 20ft, 11.7MHz, flagpole monopole feed point, over perfect ground. Adding a series 45.19pF at the feed point of the software of the 20ft monopole over perfect ground found a resonant frequency 16.56MHz.

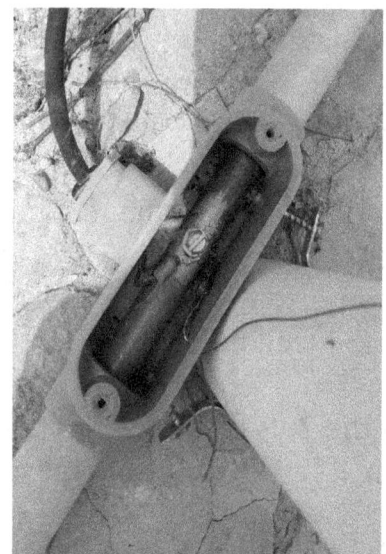

The flagpole.   The base junction feed.   The auto-tuner.

The flagpole came in four sections and is insulated from ground using PVC pipe. Electrical PVC pipe covers and supports the copper pipe capacitive ground. A short piece of coax runs from a nearby weather proof box to the PVC junction box at the flag pole base. The coax shield is connected to the center of the 10ft copper pipe inside the PVC pipe, the coax center conductor to a wire that runs to the top of the flagpole. Although the flagpole is metal, the sections did not provide good electrical connections.

By using an automatic remote tuner, at the base, with the input to the tuner fed via a broad-band (W2DU type) current choke, this bottom capacitive end loaded vertical dipole (or is it a capacitively shortened monopole?) covers 40 to 6 meters at a cost less than a multi-band trap vertical. (I run only 100 watts, higher power would require a different auto tuner, which would increase the total cost). I have an HF antenna within the rules and no radials. And it works, including DX.

137

To the neighbors who compliment me on my patriotism, I simply quote Samuel Johnson from Boswell's biography, "Patriotism is the last refuge of a scoundrel" and smile, as I admire my stealth HF vertical antenna with no buried radials.

6.6 Low Vertical Loops

In this section a vertically polarized full-wave wire loop antenna for 17M will be transmission line modeled. Full-wave loops are 1/2-wave transmission lines spread out into various shapes, often square, sometimes triangular or rectangular as in this example.

The 1/2-wave loop, as a transmission line, is shorted at the end opposite the input end. The 1/2-wave line being two 1/4-wave lines end to end inverts and inverts again bringing us back to its input. What we now have is the radiation resistance caused by a leaky shorted half-wave transmission line as an antenna with no reactance at resonance.

An alternate view of the full-wave rectangular, vertically fed loop, is two end loaded vertical dipoles, physically separated in space by the parallel transmission line between them, which is also end loading both vertical dipole ends.

When vertically fed the two horizontal wires of the interconnecting parallel transmission line have fairly equal, but opposite, currents. This results in field cancellation and very little horizontal radiation.

As a full-wave loop is stretched from a square into a rectangular (or oblong) shape, the distance between the two vertical ends, acting as vertical dipoles, will have phase shifts in space with each other. The phase shifts between the two vertical dipoles of the vertically fed rectangular loop combine to form constructive interference broadside to the loop. Those same phase shifts will produce destructive interference along the loop.

The broadside gain is dependent upon the ratio of height to length, and so is the radiation resistance at the center of a vertical end.

The more vertical the rectangular loop, and less wide, the higher the radiation resistance. Continuing this lengthening process to a taller and narrower vertical loop, eventually the loop becomes a half-wave, center fed, folded vertical dipole. As the vertically fed loop is reduced in height and horizontal length increases, becoming rectangular horizontally, the radiation resistance becomes lower.

By thinking of the vertical parts of the loop as end loaded vertical dipoles we know that the shorter the vertical part of an end loaded vertical dipole the lower the radiation resistance, the loop follows this truth.

There are many possible ratios for a rectangular loop, all with some optimum gain or feed point radiation resistance. In keeping with low profile antennas, I will model a vertical rectangular loop that is longer than tall, with the goal of a feed point radiation resistance near 50Ω for a low SWR using coaxial cable.

Since I have claimed that the rectangular loop resembles two short vertical dipoles separated by an interconnecting transmission line, I have set the stage to use calculating techniques that have been used for short, end loaded, vertical dipole antennas in previous sections.

Treating the two individual vertical sections as end loaded vertical dipoles I will select a height for close to 50Ω feed point radiation resistance. When a monopole or dipole is end loaded the relationship for the radiation resistance becomes $Rr = Ro(sin^2\Theta)$. I have decided to make the vertical dipoles at the ends of the horizontal rectangular loop 1/8-wave high (45 degrees). At 45 degrees electrical length, using an Ro of 97Ω, typical of a vertical dipole, $Rr = (97Ω)(sin^2 45 \text{ degrees}) = 48.5Ω$.

Now that 1/8-wave (0.125) has been selected for height, the horizontal length will be what's left, which is (0.375) 3/8-wave length. The loop to be modeled will have 1/8-wave vertical height and 3/8-wave horizontal length, for a radiation resistance near 50Ω.

For 17M a full-wave length at 18.1MHz will be, (984)/(18.1MHz) = 54.37ft. The height will be (54.37ft)(0.125) = 6.8ft. The length will be (54.37ft)(0.375) = 20.4ft. Or, since the length is 3 times the height, three times 6.8ft.

For a wire loop of this size and shape, with the bottom of the loop 1ft above ground, software calculated, Z = 46.5-j45.98Ω, 2.52:1 SWR at 18.1MHz.

The software model shows a bi-directional broadside gain pattern. The gain is dependent on the soil type. Higher conductivity improves the gain and lowers the elevation angle, as with any vertical antenna over real ground.

The software Rr of 46.5Ω compares reasonably well with the estimated target value of 48.5Ω. For a shorted 1/2-wave line the reactance should be, in theory, zero at resonance resulting in an estimated feed point impedance of Z = 48.5-j0Ω, for 1.03:1 SWR.

At this point I need to say that I knew the Zo of the loop antenna as a transmission line wuold not be exactly as calculated. The line does not have uniform spacing from end to end. Each end of the line is bent 90 degrees to meet the other side of the line to form the two vertical dipoles.

A software frequency sweep of the loop indicates resonance at 18.418MHz, $Z = 47.6+j0.1505\Omega$, 1.05:1 SWR, which makes the calculated target value of 48.5Ω, 1.9% higher than the software value. Thanks to software we can revise our transmission line formula (984/f) to adjust for the bent ends that form the vertical dipoles. The new formula will be $(53.37ft)(18.418MHz) = (1001/f)$ for the 1/8 by 3/8 horizontal loop.

The software calculated reactive component is $-j45.98\Omega$ at 18.1MHz, which can be tuned out with an inductor (or adjusting the size a bit). The inductor can be placed at the feed point, the center of the fed vertical dipole, *or* in the center of the opposite vertical vertical dipole. This is possible because a 1/2-wave transmission line repeats. Anything done at the unfed end will appear at the fed end.

To cancel the $-j45.98\Omega$ of capacitive reactance an inductance of $0.40\mu H$ was placed in the center of the unfed vertical end opposite the fed end. The software model input impedance becomes $Z = 46.68-j0.517$, 1.07:1 SWR.

Since inductance can be used to adjust resonance, what will happen if we decide to feed this loop at 14.1MHz instead of its design frequency of 18.1MHz? Thus far, experience has shown that when an antenna of a fixed length is operated at a lower frequency the radiation resistance lowers and there will be a capacitive reactance added to the feed point impedance.

Using $Rr = Ro(sin^2\Theta)$ and transmission line theory let's see what the calculator has to tell us about the radiation resistance of the 18.1MHz loop on 14.1MHz, The length of a wave at 14.1MHz becomes $(984/14.1) = 69.79ft$. This will result in the vertical portion being, $\Theta = (6.8ft/69.79ft)(360) = 35.08$ degrees. An estimate of the radiation resistance for an end loaded, 35.08 degree tall, vertical dipole is, $Rr = (97\Omega)(sin^2\ 35.08) = 32\Omega$.

What about the reactive component?

Using basic transmission line theory we can calculate the reactance of the less than 1/2-wave transmission line.

The characteristic impedance of a parallel wire transmission line is $Zo = 120Ln(2)(S)/d$), the spacing "S" and wire diameter "d" must be in the same units. The spacing of the parallel transmission line wires is the height of the rectangular loop vertical dipoles. For this antenna as a transmission line, using wire,
$Zo = 120Ln(2)(6.8ft)(12in/ft)/(0.0641) = 941\Omega$.

Now for the electrical length of the line at 14.1MHz. The total physical length is (6.8ft+20.4ft) = 27.2ft. The electrical length will be, (27.2ft/69.78ft)(360 degrees) = 140.3 degrees.

The total electrical length of the antenna as a transmission line at 14.1MHz is 140.3 degrees. The loop will have a capacitive reactance at 14.1MHz because it is short, not being 180 degrees at 14.1MHz.

The capacitive reactance of this 140.3 degree line at 14.1MHz will be $-jX = Zo/\tan\Theta$, in this case that will be, (941Ω)/(tan 140.3 degrees – 90 degrees) = (941Ω)/(tan 50.3 degrees) = -j781Ω.

The 90 degrees was subtracted because the vertical dipole at the end opposite the fed vertical dipole acts as a short on the transmission line. This short, 90 degrees away, is inverted to an open on the remaining parallel wire transmission line. An open on a 50.3 degree transmission line will have a capacitive reactance.

At 14.1MHz, software calculated, Z = 47.33-j739.8Ω, as compared to my calculated -j781Ω. Inserting an inductor with a reactance of +j739.8Ω, L= 739.8Ω/2π(14.1) = 8.35µH in the center of the vertical dipole opposite the fed vertical dipole, software calculated Z = 29.21-j0.6501Ω, 1.71:1 SWR, compared to my calculated 32Ω, 1.56:1 SWR estimate. Formula estimated Rr = 32Ω vs software calculated 29.21Ω. The reactance, calculated by transmission line theory was -j781Ω vs software -j739.8Ω.

Had I selected a loading inductor based on the transmission line calculated -j781Ω, the inductance would have been 8.82µH. Using that inductance in the software model the result is Z = 29.16+j24.49, 2.27:1 SWR.

Software shows bi-directional broadside gain and directivity at 14.1MHz with the inductive load, though not as much as at the design resonant frequency of 18.1MHz.

The choice of a 0.125 (1/8) by 0.375 (3/8) wave loop is a compromise to have low height, a radiation resistance close to 50Ω, with gain and directivity. Vertical rectangular loops can have directive gain and be low profile.

I had designed and modeled this 17M loop intending to load it for 20M because I didn't think a full size 20M loop would fit the space available and still be stealthy. After some walking around and measuring I found that a full size 1/8 by 3/8 20M loop would fit. Thanks to some low tree limbs I was able to support the upper wire without eye catching supports. The result was a low profile 50Ω (I adjusted the length, mostly due to the use of insulated wire), rectangular loop.

For more about antenna directivity and gain see "Antenna Gain – A Chewing Gum Theorem" in the appendix.

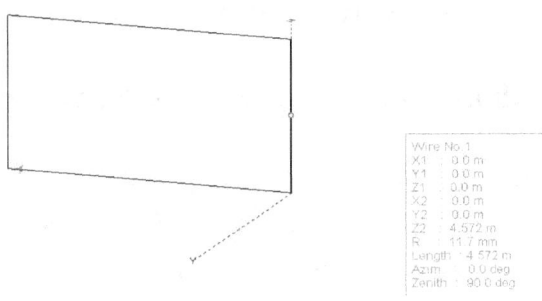

U. S. Amateur Radio Station 9DTA, circa 1920

The Amateur radio operators of the 1920's pioneered long distance radio communications on the then considered "useless" short-waves, and in my opinion created the *first* "social media."

Appendix

A1 After Words and After Thoughts

About Antenna Average Characteristic Impedance

The models presented in this book treat antennas as transmission lines with a lumped resistance to account for radiation (Rr). The antenna as transmission line model is used to calculate the lumped reactance of the antenna impedance creating a series AC circuit with a radiation resistance.

One approach for finding antenna characteristic impedance was to use the unit length distributed capacitance and inductance of a vertical conductor in the formula for a transmission line to calculate the average characteristic impedance of a linear antenna.

Around 1914, Professor George William Osborn Howe derived estimates of the inductance and capacitance, per unit length, of a vertical conductor. Using his inductance and capacitance values in the formula for a transmission line, $Zo = \sqrt{(L/C)}$, the characteristic impedance of a grounded vertical conductor is $Za = 60[Ln(2h/d) -1]$, where "h" is the height and "d" is the diameter of the vertical conductor, in the same units. The Howe formula has been used throughout this book.

Another relationship for the characteristic impedance of a monopole was provided by Carl (Charles) Steinmetz. His formula works out to be, $Za = [60Ln(4L/d)] -104$. Here "L" is the length of the monopole and "d" is diameter, both in the same units.

Using a 10ft (3.048m) conduit, $d = 24mm$, modeled in software, over perfect ground, $Za = 275.2\Omega$ by finding the capacitive reactance at half the resonant frequency.

For the Howe TL derived formula, $Za = 272\Omega$, 1.2% lower than software. The Steinmetz $Za = 270\Omega$, 1.9% lower than software.

As for the change of the formula for a horizontal dipole above ground to -0.75 rather than -1, is my average adjustment based on a formula given by Robert Dome, W2WAM, in his article "Impedance of Short Horizontal Dipoles", QST January 1976, page 33. He credits that formula to "Reference Data for Radio Engineers" 3rd edition 1949, which combines the free space Za of the dipole with its Zo as a parallel wire transmission line with the ground image.

In one source an author states he would rather call the antenna characteristic impedance the "wave impedance" of the antenna, another author called it the "surge impedance".

By the 1970's the transmission line equivalent for an antenna lost out to the method of moments software (MoM) and the calculation ability of computers. By the 1980s MoM software was running on personal computers.

Antenna Modeling Software

MMANA-GAL (software #1) //hamsoft.ca/pages/mmana-gal.php
MininecPro (software #2) www.blackcatsystems.com/software/mininec.html
AN-SOF100 (software #3) www.antennasimulator.com
All three are no cost versions. MMANA-GAL, software #1, was used throughout this book to validate the antenna as transmission line models.

Don't Let Coax Ruin Your Antenna

All the models, both antenna as a transmission line and software, place the driving source at the feed point.

In the real world a feed line (aka feeder) is used. All coaxial lines need to be isolated from the antenna. If not isolated, antenna current can flow on the outside of the coax shield and become part of the antenna.

All the examples in this book, fed by coax, need a good quality RF choke or current balun at the feed point to prevent coax outer shield current flow.

All my measured values were taken using a W2DU type ferrite bead choke.

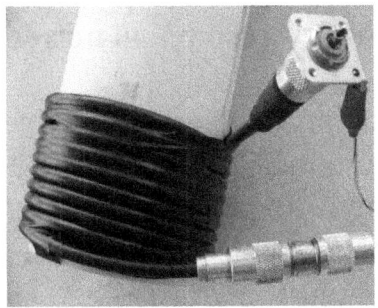 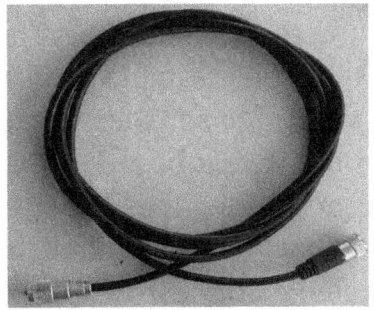

Coiled coax chokes are not broadband. The inductance and capacitance of the coax shield forms a parallel circuit. Either side of resonance the coax coil will have a capacitive or inductive reactance. The reactance will add to the antenna feed point impedance and change the antenna natural resonance. (see the "Trap Antenna" section)

Real World Antennas

Models assume horizontal radiators are linear. In the real world wires sag. Models also assume no insulation on conductors and free space dielectric values. All the models in this book assume bare wire. Insulation changes the wire capacitance and its electrical length. Expect to adjust wire lengths.

Also, modeling cannot know the antenna environment is not ideal. Real antennas will have nearby objects, supporting structures and insulators. As much as the antenna as transmission line model and computer software agree they are both dumb to the realities of the actual environment and are based on ideal conditions.

The Loss of the Counterpoise

During the Trans-Atlantic tests in 1923 the first two-way Amateur Radio contacts were made on short wave frequencies. Short wave in the early 1920's would have been 100 to 200 meters wavelength (1.5 to 3MHz).

One of the US stations contacting French station 8AB was operated by John L. Reinartz, US 1QP/1XAM using what today we might see as an inverted L antenna operated against an elevated counterpoise.

The antenna was made up of three cages arranged in a C shape. A drawing of the Reinartz antenna was included in the January 1924 QST article about the success of the transatlantic tests and the equipment used, on both sides of the Atlantic. Contacts were reported on 110 meters (about 2.7MHz) and 115 meters (2.6MHz). French station 8AB also used an antenna with a counterpoise.

Counterpoise Conductors Can be Bent

Although the conductors for the counterpoise examples in this book were linear, they can be bent to be more compact.

The counterpoise conductors should be bent in a way to retain mirror image symmetry so the currents in the two cancel.

This is where the software has a great application, finding the relative gains of monopoles with different bent counterpoise configurations.

In an effort to keep the bent counterpoise simple to apply and install, I restricted myself to one 90 degree bend for each of the two counterpoise conductors. This restriction created rectilinear cases, rectangles and square.

With this simple limitation, software showed that the best counterpoise shape was a "U", where each of the three sides are the same length and had one 90 degree bend in each of the two counterpoise conductors.

This U configuration improved the gain of the counterpoise/monopole combination by almost 1.25dB over the linear counterpoise conductors.

The best U shape within my restrictions, according to software, was an open square. The connection point being the center of the bottom side of the U shape.

To find the lengths of the three sides of the open square U, where "L" is length of one linear counterpoise wire and "X" is the sides and bottom lengths of the square U, $X = (2/3)(L)$.

According to software models, in some cases of bent counterpoise wires there is a small increase in the gain of the monopole over straight wires.

Which Measuring Stick?

Metric vs English. Metric is really much easier, and more systematic (I have 10 fingers, not 12!). So, what to do when writing this book? Software #1 uses Metric, which I like. But feet and inches are common in the U.S., certain antenna materials, such as conduit lengths, are in feet. Therefore I have used some of each in this book.

Here is the conversion number I use, 0.3048. To get from feet to meters multiply by this number. To get from meters to feet, divide by this number.

Another Use for a Geometric Mean

The characteristic impedance of an unknown (or home brew) transmission line can be found using the geometric mean of the measured open and shorted reactance.

The characteristic impedance is found from, $\mathbf{Zo = \sqrt{(Zoc)(Zsc)}}$, where Zoc is the transmission line open impedance, and Zsc is the impedance with a short circuit. This works for any length line that is not an exact wavelength or multiple of a wavelength.

Calculating SWR with a Complex Load - No Smith Chart Required

For a *purely resistive* load, the SWR can be easily calculated by R/Zo for R greater than Zo or Zo/R for Zo greater than R.

It is possible to find the *SWR for a complex load* of R + jX or R- jX with a calculator. There is no need for a Smith chart or computer software.

Step 1, (R + jX)/(Zo) = r + jx. This step is called "normalizing" the complex impedance in terms of the reference impedance, Zo, the transmission line characteristic impedance, often 50Ω. (Anyone who has used a Smith Chart has done this step before).

Step 2, find $b = [(jx)^2 +1)/(r)] + r$.

Step 3, calculate $SWR = [b + \sqrt{[(b^2 - 4)/(2)]}$.

More About Antenna Bandwidth

If we apply the RLC series circuit bandwidth definition of -3dB to an antenna, using the two -3dB frequencies to define antenna bandwidth, that would at least be following a standard.

In the section "Antenna Reactance, Resistance, Bandwidth and Q" I represented the antenna model as a series RLC circuit, where R was the radiation resistance. This is a simplified ideal model for the antenna impedance.

For an antenna, the radiation resistance is not a fixed, constant value without regard to frequency. It is not a resistor in the sense of a resistor found as a circuit component.

Radiation resistance varies with frequency unlike the resistance in the ideal RLC series circuit. The variation in radiation resistance with frequency follows trigonometric sine and cosine functions.

The reactance of an antenna does not follow the relationships $+jX = 2\pi fL$ or $-jX = 1/(2\pi fC)$ as in the case of a component RLC circuit. Antenna reactance follows transmission line relationships, $+jX = (Za)(\tan\Theta)$ and $-jX = (Za)/(\tan\Theta)$, which are also frequency (wavelength) dependent.

Since the radiation resistance is changing along with reactance as the frequency is varied either side of resonance, finding the -3dB points is not quite like the simple component RLC model when applied to an antenna, but let's try it anyway.

I used MMANA-GAL software to model a 20ft (6.096m) wire, center fed dipole, in free space. Software found the resonant frequency to be 23.832MHz, 1.44 SWR. The two -3dB frequencies (where Rr = jX) found were 22.783MHz, 2.97 SWR and 25.292MHz, 3.81 SWR.

The -3dB bandwidth of this wire dipole = (25.292MHz – 22.783MHz) = 2.509MHz. The Q = (23.832MHz/2.509MHz) = 9.5.

Modeling a 10ft (3.048m) wire monopole over perfect ground, software found the resonant frequency to be 23.833MHz, 1.39 SWR. The two -3dB frequencies were 22.783MHz, 2.44 SWR and 25.294MHz, 2.49 SWR.

The -3dB bandwidth = (25.294MHz – 22.783MHz) = 2.511MHz. The Q = (23.833/2.511) = 9.5.

By using the -3dB method, the bandwidth and Q of the wire dipole in free space and the wire monopole with its ideal image are the same. which validates the idea that dipoles are two monopoles in series.

Now look at the SWR values at half power frequencies for the dipole compared to the monopole. Though the SWR values at resonance are very close for the two antennas, the SWR at the -3dB points are quite different between the two.

As we know, the diameter of the antenna conductor has a direct effect on bandwidth, which is why I modeled a wire dipole and a wire monopole so the conductor diameters (r = 0.8mm) would be the same. Doing so gave the same Q for the dipole and the monopole using -3dB values.

Now for a 10ft (3.048m) conduit (radius = 10.5mm) monopole over a perfect ground to find the resonant frequency and the two -3dB frequencies.

For a 10ft (3.048m) conduit monopole over perfect ground, software found the resonant frequency to be 23.421MHz, 1.39 SWR. The two -3dB frequencies were 21.955MHz, 2.48 SWR and 26.100MHz, 2.72 SWR.

The -3dB bandwidth = (26.100MHz – 21.955MHz) = 4.145MHz. The Q = (23.421MHz/4.145MHz) = 5.7.

The bandwidth increased and the Q became smaller. Remember, fat antennas have a wider bandwidth than thin? This is because the antenna reactance as the frequency changes from resonance, is dependent on the characteristic impedance of the antenna Za. and the lower Za (fat conductor). The lower the -jX or +jX variations of the antenna are either side of resonance and the wider the bandwidth.

If a -3dB bandwidth isn't used but a 2.5:1 SWR bandwidth is used instead, is that a reasonable "standard" for antenna bandwidth?

To answer the 2.5:1 SWR bandwidth question, I once again turned to software. This time, instead of looking for two -3dB (Rr = jX) frequencies I looked for the two frequencies either side of resonance at which the SWR was 2.5:1 for the same wire monopole and dipole antennas.

For the 20ft (6.096m) wire dipole in free space, software found the two 2.5:1 SWR frequencies to be 22.965MHz and 24.700MHz, for a 2.5:1 SWR bandwidth of 1.74MHz.

For the 10ft (3.048m) wire monopole over perfect ground, software found 22.750MHz and 25.300MHz for the two 2.5:1 SWR frequencies and a 2.55MHz bandwidth.

This is interesting, a dipole being two monopoles in series, yet the 2.5:1 bandwidths are not the same. Does it seem reasonable that the dipole bandwidth would be less than the monopole for the same resonant frequency and conductor size? It is if SWR bandwidth is used.

What if the antenna had a 2.5:1 SWR at resonance? Would that mean the antenna has no bandwidth? What if an antenna had an SWR of 2.7:1 at resonance, a negative bandwidth?

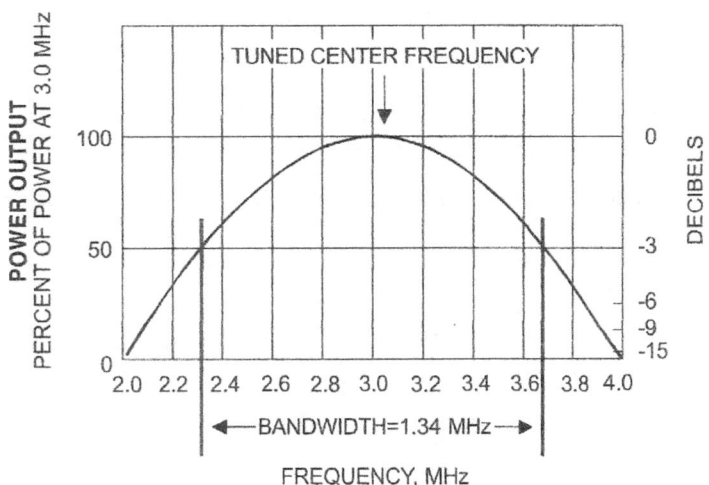

Here is an antenna bandwidth graph from MCRP 3-40.3C, U.S. Marine Corps "Antenna Handbook." This graph shows the bandwidth related to the half power points. It makes no reference to the type of antenna, monopole or dipole, or SWR, just the resonant frequency and the bandwidth at the *-3dB* points.

148

The Q of this antenna is 3MHz/1.34MHz = 2.24. With a Q that low the antenna would need to have a low Za. A low Za would require fat antenna conductor(s).

All the previous discussion of -3dB and SWR bandwidth calculations put the source at the antenna feed point. In most cases it is likely the SWR bandwidth will be made at the other end of some length of feed line. This makes the SWR bandwidth a system (antenna & coax), rather than an antenna bandwidth. Due to the usually unknown length and loss of the transmission line, the actual bandwidth of the system can be greater at the transmission line input than at the antenna. Transmission line loss will give the antenna the appearance of a wider bandwidth than it actually has.

As Les Moxon, G6XN, put it in his book HF Antennas for All Locations, "SWR bandwidth, though often used as the main criterion of "bandwidth", it is of little *intrinsic* importance." (his italics)

Why is Voltage Sometimes "E" and Sometimes "V"?
And Why is Current "I"?

As technology evolves, grasping the new often involves comparisons to the old and already understood.

The application of damping in mechanical systems to electrical is an example. Eventually damping was replaced by Q and bandwidth in radio electronics.

As the evolution of the understanding of electric technology moved from static charges to moving charges, the moving charges, like fluids, were considered to "flow." In a fluid system there was a rate concept called "intensity" of flow.

The "intensity" of electric current flow can be seen in an incandescent bulb. As the current flow increases through the bulb filament the radiated light becomes more "intense." From the intensity of electric current flow came "I" for a flow rate. Many an impoverished amateur radio operator used an electric light bulb, at a current location in an antenna system, for a visual indication of current flow ("tune for maximum glow, minimum smoke"), because an RF ammeter was beyond their finances.

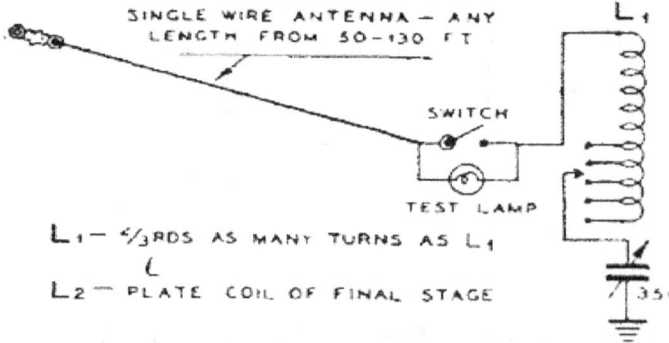

Here is an example of placing a bulb in series with the antenna wire to observe the "intensity" of RF current flow. Once tuned, the switch was closed to bypass the bulb resistance. ("Jones Antenna Handbook", 1937)

For voltages, a source of electric charges was called an "electromotive force" (EMF). Sources were considered "E" and the units were in volts.

As current flows through a resistor, R, the work being done creates a "voltage drop", this voltage is not like a "source" it is a consequence of current flowing and work being done, identified as "V" not "E".

Using a strict application in electric circuits, E would be used for voltage source(s) to distinguish them from V, voltage drop(s) across loads.

Then along came alternating currents, and with active devices (tubes and later transistors), we had circuits that contain both DC and AC together.

Now we needed to differentiate the AC voltage and current from the DC voltage and current. The most common way to do this is to use the capital letters (E if a source, or V if a drop, R and I) for DC and use lower case letters (e if a source, or v if a drop, r and i) for AC. This is why, in AC circuits, the imaginary reactive oppositions use +j or -j rather than "i" used for $\sqrt{-1}$ in mathematics.

When both AC and DC exist together in a circuit it is common to say that the AC is "superimposed" on the DC. The superposition theorem has wide application in electricity and electronics.

The superposition theorem was applied to an antenna in the section on horizontal dipoles and the ground image related to constructive and destructive interference of the wave fields. Superposition is also used to determine the standing wave when forward and reflected waves combine.

James Clerk Maxwell, Meet Thomas Crapper

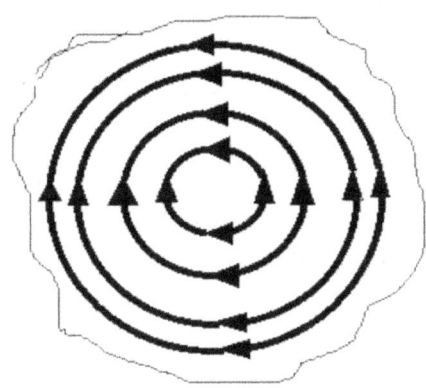

James Maxwell   **Circulation**   **Thomas Crapper**

Even in the world of vector theory for electric and magnetic fields we find flow concepts. Maxwell studied heat flow and electromagnetism.

There are terms used in the vector calculus of fields that are related to flow concepts.

The term "grad", which is short for gradient, is really about slope, a rate of change. Water in a V shaped stream flows faster at the bottom than at the top.

Then there is "div" which is short for divergence. Electric fields can diverge or converge. Water diverges from a hose. Streams converge into rivers.

The term "curl", also called circulation, is applied to the rate of field rotation. Magnetic fields encircle current carrying wires. Whirlpools in water can have rotation due to flow. So can the coffee in your cup as you stir the contents.

You see curl (rotation) every time you flush a toilet, there is gradient in the slope of the sides of the toilet bowl.

You don't usually get to see the divergence of the flush as it goes down the drain pipe, unfortunately (or maybe not?). Those flushes converge at the sewer plant.

The Crapper Antenna

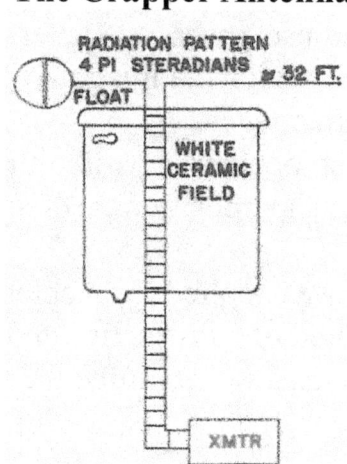

With the loss of copper float balls to plastic, this antenna fell out of favour.

Radio Explained

"You see, wire telegraph is a kind of a very, very long cat. You pull his tail in New York and his head is meowing in Los Angeles. Do you understand this? And radio operates exactly the same way: you send signals here, they receive them there. The only difference is that there is no cat."

Albert Einstein

A dB Story - Pigs is Pigs?

A farmer was telling a young electrical engineer the reason he raised a breed of hogs was because the sows of that breed had twice as many pigs as another breed.

The electrical engineer nodded knowingly, and muttered, "yes, 3dB more."

John Lenahan, KØRW

A2 Free Space – The Transmission Line of the Universe

Throughout the various examples in this book the characteristic impedance of antennas, parallel wire transmission lines and single-wire transmission lines have been calculated.

What is characteristic impedance? Impedance implies an Ohmic opposition, and all the formulas for characteristic impedance were in units of Ohms.

Characteristic impedance, at its most basic, is a description of the ratio of voltage to the current, V/I (or v/i).

The impedance of the transmission line is formed by characteristics of the line, the inductance and capacitance per unit length, resulting in a characteristic impedance in Ohms.

All of the characteristic impedance formulas used so far contain conductor length and diameter or diameter and spacing of conductors.

Any transmission line, regardless of characteristic impedance, acts as a wave guide. The characteristic impedance of a transmission line sets the wave impedance, the ratio of the wave voltage to wave current, as power travels from input down the line to the output.

When a wave is guided, the electric and magnetic fields of the wave will have a ratio of the electric field divided by the magnetic field, which is the wave impedance in Ohms.

In an ideal transmission line, one with no loss, the power being guided is moving along the line in the electric and magnetic fields as they flow (propagate).

Anyone who has used a microwave oven knows that electric waves have power.

A transmission line provides storage in the capacitance and inductance per unit length as it guides the wave. Capacitance between the conductors, inductance along the conductors (the power is in the electric and magnetic fields).

In an ideal wire transmission line the characteristic impedance $Zo = \sqrt{(L/C)}$, where L and C are the inductance and capacitance of the line per unit length.

The fields within the line are self contained. The electric field extends from one conductor surface to the other surface and the magnetic field encircles the conductors in a ratio that is set by the characteristic impedance of the transmission line guiding the wave.

Within the ideal transmission line the electric and magnetic fields are uniform and balanced so they do not radiate.

To radiate, the balance has to be disturbed. This led to the early approximation, used in this book, that the antenna was a spread out transmission line and that the antenna has a characteristic impedance, an antenna wave impedance, which I called Za.

The antenna, unlike the usual transmission line, has a very important function - to get those very happily balanced electric and magnetic waves of the feed line to head out into space, to change guided energy in the feed line into a radiated electromagnetic field.

If radiation is to occur, space must be able to transport electric energy in the form of fields. This implies that space must have a characteristic impedance, a wave impedance, and is capable of guiding energy for transport as electric and magnetic fields like any other transmission line.

Since space has no conductors for inductance or to form capacitance, how can space have an impedance and transport the radiated energy of our transmitted signal (or microwave oven)?

Space has built into it its own capacitance and inductance per unit length. These values were measured and known before Maxwell did his breakthrough work. In fact, he used the known values of capacitance and inductance of space to prove that all electromagnetic waves travel at the same speed, the speed of light in space.

As an electromagnetic wave travels through space the wave will have a ratio of electric field to magnetic field that describes an impedance in Ohms, known as the wave impedance of free space.

The inductance of free space is $4\pi10^{-7}$ Henry/meter. The capacitance of free space is $10^{-9}/36\pi$ Farads/ meter.

If you put these in the formula for characteristic impedance, $\sqrt{(L/C)}$, the result is Ohms of impedance, 376.99Ω, usually rounded to 377Ω, or 120πΩ.

This free space impedance is called the intrinsic wave impedance of free space, intrinsic because it is built into space.

The antenna as a transmission line is transforming the wave impedance it has, to the wave impedance of free space, and vice versa. It is assisting the closed lines of electric and magnetic fields of the transmission line to spread out into the free space impedance so the wave can expand into a plane wave and radiate out into space.

The electric and magnetic plane waves then can be intercepted by another antenna and induce power into a load at its feed point. This process is like the photons from the sun on a solar cell moving electrons to provide power to a load.

The antenna at resonance is a special sort of wave transformer, using its wave impedance to convert the wave impedance of free space to a feed point impedance.

The beauty of free space impedance is it has no loss due to resistance or conductance, perfect for energy transport.

If you place a sphere around any antenna at a distance where the wave has flattened out into a plane wave, and sum the power passing everywhere through that imaginary spherical surface, it will always be the same, no matter the radius of the imaginary sphere or antenna pattern.

What changes as the sphere becomes larger (think greater distance from the antenna) is the amount of power per unit area becomes less. The power spreads over more surface area, but total power passing through the entire sphere, no matter the size, remains the same, regardless of distance.

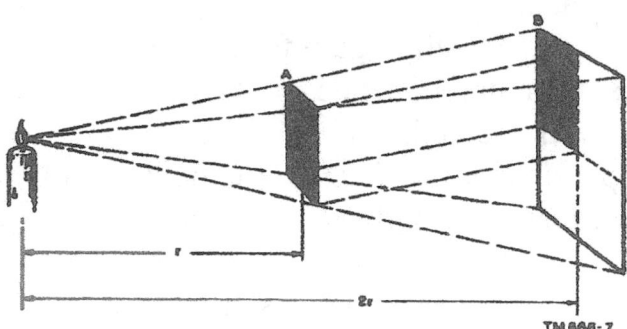

Figure 12. How light waves decrease in intensity as the square of the distance.

There is no loss of power in free space, just an increasing distribution of the total energy over a larger and larger area.

When you look at night sky objects your optical sensors can "see" stars because space is a transmission line. All the light from those stars have their power brought to you in a ratio of electric to magnetic fields set by the capacitance and inductance of free space. The energy of starlight is being delivered to your eyes by the loss-less transmission line of free space.

The transport of all radiated electromagnetic energy relies on the universal transmission line that is free space.

Whether you are working that "new one", heating food in the microwave, or star gazing, it's all possible thanks to the impedance of free space.

A3 Antenna Gain – A Chewing Gum Theorem

What is antenna gain? Gain means an antenna has a greater field strength in some directions than others.

For an antenna to have gain in some direction(s), it must have less (negative gain in the world of decibels) in other directions.

With gain comes directivity, with directivity comes gain.

All antennas have the same total average gain. Don't believe such a crazy assertion?

I have an experiment for you! All you need is a blob of chewing gum (already chewed), a straight clear glass partially filled with water and a toothpick (I like a small glass for this, and chew less gum). Mark the level of the water on the side of the glass.

Roll your blob of chewing gum into the best, most spherical ball you can make. Stick the toothpick into it and holding the toothpick's opposite end, submerge the ball into the water. If water goes over the edge of the glass you had too much water and need to start over with less water. You need a water line mark without the gum as a reference.

When your ball of gum is submerged in the water, mark a new line for the water level with the ball in the water.

Remove your toth pick and ball of gum from the water, and make it into a doughnut shape by pressing on opposite sides of the ball. Re-submerge this new shape into the water. The water will rise to the same height as it did for the round ball shape.

This doughnut approximates the shape of the field of a dipole in free space. The gum is now more in some directions and less in others, but as far as filling space it is the same.

Go crazy and make other shapes (no voids or hollow spaces are allowed, fields don't work that way), the volume of water displaced when your shape is immersed will rise to the same level.

Your original nice round ball of gum is similar to an isotropic radiator (which has a power gain of +1dB, or 0dBi), radiating equally in all 360 degrees. An isotropic radiator would illuminate a sphere around it uniformly, just as the ball of gum presses equally on the volume of water in all directions. Antenna gains are often compared to a theoretical isotropic antenna in dBi.

The field patterns you see in software are telling you where the gum was "squished" to poke more radiation in some direction, less in others.

You can get a "visual" for why the ends of the dipole have very little radiation and the dipole gets that doughnut shape (broad-side radiation pattern).

Hold a pencil out in front of you at arms length looking at the side. The pencil is radiating light. It is reflected light, but it is radiating, which is why you can see it (unless you are in a dark room).

Now slowly rotate the pencil from your broadside view to an end view. Notice the amount of pencil (and reflected light) seen reduces and reduces until the end is straight at you, which has very little surface to radiate light.

You have seen with your eyes what would be seen by a receiving antenna at various angles to a radiating dipole antenna exploring the antenna field strength in free space.

In the world of antenna fields your eyes saw variations in antenna aperture as you rotated the pencil from side view to end view. Antenna gain and direction are related to antenna aperture. If that pencil had been a light source like a dipole, it would have illuminated a surrounding sphere more from the middle than at the ends.

If the length of the illuminating pencil were reduced (or half of it covered over, as in the case of a shortened dipole), the illumination of the sphere would be reduced.

To obtain the same level of illumination from the shorter light source it would have to increase brightness.

To increase the brightness the emitted power would have to be increased. This would require the current through the illuminating source to increase, because current does the work in electric circuits, current would have to increase.

According to Ohm's law, for current to increase, the resistance of the light source must decrease.

To produce similar illumination of the free space sphere by an electromagnetic field, the current in a short antenna must increase, leading to a reduced radiation resistance of the antenna. This causes a short dipole (or monopole, which is half a dipole) to have a lower radiation resistance than a resonant antenna.

When measuring the field strength of an antenna it is important to be far enough from it to be in the "far-field."

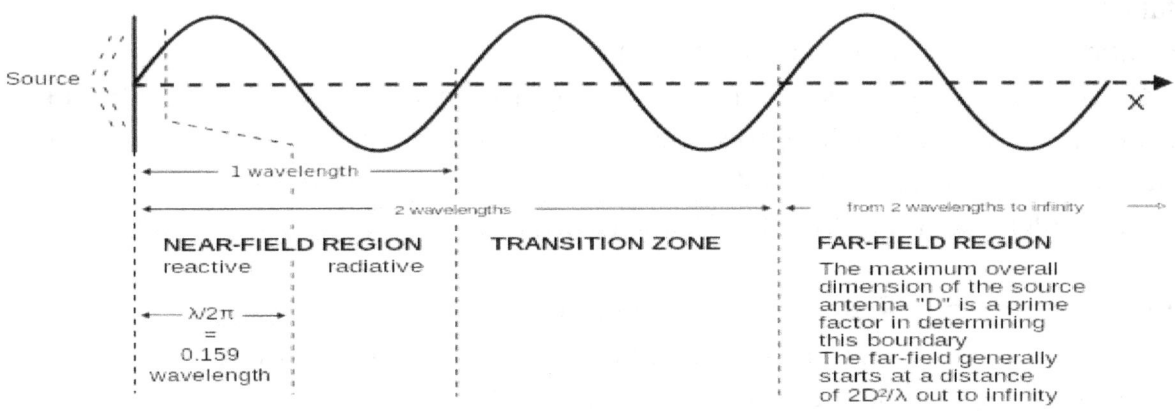

As can be seen in this graphic the distance for the far field of an antenna should be "from 2 wavelengths to infinity" as a general guide.

The included formula, for a particular antenna, may be a lesser distance than the general guideline. The calculated distance should be exceeded to ensure the measurements are in the far-field.

A4 Formula List

Average Characteristic Impedance of an Antenna
Monopole: $Za = 60[(\ln 2L/d)-1)]$
Dipole in Free Space and Vertical Dipole: $Za = 120[(\ln 2L/d)-1)]$
Horizontal dipole (Above Ground): $Za = 120[(\ln 2L/d)-0.75)]$
Where L is length, d is diameter, same units.

Characteristic Impedance of Wire Transmission Lines
Parallel Wire: $Zo = 120 \, Ln(2s/d)$
Single Wire Over Ground: $Zo = 60 \, Ln(4h/d)$, where s is spacing, d is diameter, and h is height, in the same units.

Transmission Lines and Antennas of 1/4-Wave

A 1/4-wave transformer: $R_{out} = (Zo)^2/R_{in}$, or $-jX_{out} = (Zo)^2/+jX_{in}$, and $+jX_{out} = (Zo)^2/-jX_{in}$
A shorted, ideal, 1/4-wave line appears open and resembles a parallel resonant circuit.
An open, ideal, 1/4-wave line appears like a short and resembles a series resonant circuit.

Transmission Lines and Antennas of Less Than 1/4-Wave

An open less than 1/4-wave transmission line or antenna will have a capacitve reactance.
For an open transmission line, $-jX = Zo/(\tan\Theta)$.
For an open antenna, $-jX = Za/(\tan\Theta)$.
A shorted less than 1/4-wave transmission line or antenna will have an inductive reactance.
For a shorted transmission line, $+jX = Zo(\tan\Theta)$.
For a shorted antenna (or open end antenna longer then 1/4-wave), $+jX = Za(\tan\Theta)$.
Where Zo is the characteristic impedance of the transmission line, and Za is the average characteristic impedance of the antenna.
And Θ is the length of the line or antenna in degrees.

Estimating Radiation Resistance of Antennas

For the case of no loading or base inductive, $Rr = Ro\ [(1-\cos\Theta)^2)/(\sin\Theta)^2]$
For the case of capacitive end loading, $Rr = Ro(\sin\Theta)^2$
Where Θ is the length of the antenna in degrees.
To estimate the radiation resistance with a loading coil at other than the base,
$Rr = Ro[(\sin^2\Theta_1) + (1- \sin^2\Theta_2)]$, where Θ_1 is the length in degrees below the coil, and Θ_2 is 90 degrees minus the length above the coil in degrees.
For a monopole, $Ro = 36.6\Omega$.
For a vertical dipole close to ground, $Ro = 97\Omega$
For a dipole in free space, $Ro = 73.2\Omega$.
For a dipole above ground, a graph is used to find Ro.
These relationships are for antennas shorter than resonance, where monopole resonant length in feet is 234/frequency (in MHz) and a dipole is 468/frequency (MHz).
The length of a wave in free space, 984/f (in MHz) feet.
The feed point resistance of an off-center fed dipole is, $R = Ro/(\cos\Theta)^2$ where Θ is the distance in degrees from the center of the dipole to the feed point. Or, $R = Ro/(\sin\Theta)^2$ where Θ is degrees from the dipole end to the feed point.

Capacitance of Some Geometric Forms

The capacitance, in pF of a solid disk, a sphere and a cylinder, where diameter d is in centimeters. (These are true for free space and top-loaded monopoles over infinite perfect ground. Over finite ground systems expect about half the calculated capacitance.)

Disk, C = 0.345d
Sphere, C = 0.556d
Cylinder, C = 0.802d, for diameter "d" equal height.

AC Circuit Values

Reactance of a capacitor, $-jX = 1/(2\pi fC)$, where $-jX$ is in Ohms, f is frequency and C is capacitance in Farads.

Reactance of an inductor, $+jX = (2\pi fL)$, where $+jX$ is in Ohms, f is frequency and L is inductance Henrys.

Bandwidth of a series RLC circuit, $BW = Fr/Q$, where Fr is the resonant frequency and $Q = X/R$.

If you need to "brush-up" on AC circuit theory I recommend you do an internet search for U.S. Army TM 11-681 Electrical Fundamentals (Alternating Current). This is a well written training manual and has many worked examples.

From the 1912 booklet "Wireless Course in 20 Lessons"
by Gernsback, Secor and Lescarboura

The End

Thank You Susie

www.ingramcontent.com/pod-product-compliance
Lightning Source LLC
Chambersburg PA
CBHW081125170526
45165CB00008B/2554